ESSENTIAL SQA EXAM PRACTICE

HIGHER BIOLOGY
Practice Questions & Exam Papers

QUESTIONS & PAPERS

Practise **100+ questions** covering every question type and topic

Complete **2 practice papers** that mirror the real SQA exams

Billy Dickson
Graham Moffat

HODDER GIBSON
AN HACHETTE UK COMPANY

Orders: please contact Bookpoint Ltd, 130 Park Drive, Milton Park, Abingdon, Oxon OX14 4SE. Telephone: (44) 01235 827827. Fax: (44) 01235 400401. Email education@bookpoint.co.uk. Lines are open from 9 a.m. to 5 p.m., Monday to Friday, with a 24-hour message answering service. Visit our website at www.hoddereducation.co.uk. If you have queries or questions that aren't about an order you can contact us at hoddergibson@hodder.co.uk

© Billy Dickson and Graham Moffat 2019

First published in 2019 by
Hodder Gibson, an imprint of Hodder Education
An Hachette UK Company
211 St Vincent Street
Glasgow, G2 5QY

Impression number	5	4	3	2	1
Year	2023	2022	2021	2020	2019

Illustrations by Aptara Inc.

Typeset in India by Aptara Inc.

Printed and bound by CPI Group (UK) Ltd, Croydon CR0 4YY

A catalogue record for this title is available from the British Library.

ISBN: 978 1 5104 7174 0

SCOTLAND EXCEL

We are an approved supplier on the Scotland Excel framework.

Schools can find us on their procurement system as:

Hodder & Stoughton Limited t/a Hodder Gibson.

CONTENTS

INTRODUCTION

Higher Biology

The assessment materials included in this book are designed to provide practice and to support revision for the Higher Biology Course examination, which is worth 80% of the final grade for this course.

The materials are provided in two sections.

1 Practice Questions
2 Practice Exams: Practice Exam A and Practice Exam B

Together, these give overall and comprehensive coverage of the assessment of

▶ Knowledge and its Application: demonstrating knowledge and understanding (dKU) and applying knowledge and understanding (aKU)

▶ Skills of Scientific Inquiry (SSI): planning, selecting, presenting, processing, predicting, concluding, evaluating.

(See the section on student margins on page v for the specific requirements of these types of assessment.)

The Higher Biology course is split into three areas of study: 1 DNA and the genome; 2 Metabolism and survival; 3 Sustainability and interdependence. These areas are further divided into Key Areas, which we have numbered 1.1, 1.2, … , 2.1, 2.2, … , 3.1, 3.2, … , etc. Please refer to the course specification document, which can be found at https://www.sqa.org.uk/files_ccc/HigherCourseSpecBiology.pdf, for full descriptions of the requirements of each Key Area. This is an important tool in your revision.

Practice Questions

The practice questions are arranged in sets to reflect the different types of question which are used in the Higher Biology examination, as shown in the grid below. There are 300 marks worth of questions.

1 – Multiple choice	2 – Structured and extended response questions				
	Short answer	Data handling	Experimental	Mini extended response	Full extended response
Objective items testing knowledge and understanding (KU) and some testing SSI	Structured questions testing KU within Key Areas	Structured questions testing data-handling skills	Structured questions testing experimental skills	Extended writing questions testing related KU from a Key Area. Usually worth 4–6 marks	Extended writing questions testing related KU from a Key Area. Usually worth 7–10 marks
		(SSI: planning, selecting, presenting, processing, predicting, concluding, evaluating)			

≫ HOW TO ANSWER

Each set of questions starts with a short commentary which describes the question type and gives some support in answering them.

Practice Questions Key Area index grid

The Practice Questions index grid on page vii shows the pattern of coverage of the knowledge in the Key Areas and the skills across the practice questions.

After working on questions from Key Areas across an area of study, you might want to use the check boxes to assess your progress. We suggest marking like this [–] if you are having difficulty (less than half marks), like this [+] if you have done further work and are more comfortable (more than half marks), and like this [*] if you are confident you have learned and understood an entire area of study (nearly full marks). Alternatively, you could use a 'traffic light' system using colours – red for 'not understood', orange for 'more work needed' and green for 'fully understood'. **If you continue to struggle with a set of Key Area questions, you should see your teacher for extra help.**

Practice Exams

Each Practice Exam has been carefully assembled to be very similar to a typical Higher examination question paper. Your Higher examination has 120 marks and is divided into two papers.

Paper 1 – Objective test, which contains 25 multiple choice items worth one mark each and totalling 25 marks altogether.

Paper 2 – Structured questions, which also contain extended response questions, totalling 95 marks altogether.

In each Practice Exam, the marks are distributed evenly across the three Key Areas of the course and about 90 marks are for the demonstration and application of knowledge. The remaining marks are for the application of skills of scientific inquiry.

In each Practice Exam, 70% of the marks are set at the standard of Grade C and the remaining 30% are more difficult marks set at the standard for Grade A. We have attempted to construct each Practice Exam to represent the typical range of demand in a Higher Biology paper.

Grading

The two Practice Exams are designed to be equally demanding and to reflect the National Standard of a typical SQA paper. Each exam has 120 marks – if you score 60 marks that's a C pass. You will need about 72 marks for a B pass and about 84 marks for an A. These figures are a rough guide only.

Timing

If you are attempting a full Practice Exam, limit yourself to **3 hours** to complete it. Get someone to time you! You should take no more than 40 minutes for Paper 1 and no more than 2 hours and 20 minutes for Paper 2.

Practice Exam Key Area index grids

The Practice Exam index grids on pages viii–ix show the pattern of coverage of the knowledge in the Key Areas and the skills across the two papers. We have provided a marks total for each Key Area and each Skill. Scoring more than half of these marks suggests you have a good grasp of the content of that specific Key Area or Skill. You could use the check boxes to record your total mark for each area of study.

Student margins

All question pages have margins. The margins have a key to each question to show what is being tested, as shown in the table below.

	Key	Meaning
Knowledge and its Application	Demonstrating KU	Demonstrating knowledge and understanding of biology
	Applying KU	Applying knowledge of biology to new situations, interpreting information and solving problems
Skills of Scientific Inquiry (SSI)	Planning	Planning or designing experiments to test given hypotheses
	Selecting	Selecting information from a variety of sources
	Presenting	Presenting information appropriately in a variety of forms
	Processing	Processing information using calculations and units where appropriate
	Predicting	Making predictions and generalisations based on evidence
	Concluding	Drawing valid conclusions and giving explanations supported by evidence
	Evaluating	Suggesting improvements to experiments and investigations

Techniques

There are **six laboratory techniques** with which you should be familiar for your Higher Biology **Exam**. You might even use one of these for your **Assignment**. The grid below shows the questions across this book which are related to these techniques. If you are not sure about any of the techniques or the questions referenced, you should ask your teacher for some advice.

Technique	Practice Questions					Practice Exams			
						Exam A		Exam B	
		Structured and extended response				Paper 1	Paper 2	Paper 1	Paper 2
	Multiple choice	Short answer	Data	Experimental	Extended response	Multiple choice	Structured and extended response	Multiple choice	Structured and extended response
1 Using chromatography	60						14		
2 Using gel electrophoresis	4		1						
3 Altering enzyme reaction rates									7
4 Using a respirometer	38		2						
5 Measuring metabolic rate	27	2							
6 Using a spectroscope	40					17			

Using the questions and practice exams

We recommend working between attempting questions or papers and studying the answers (see below).

Where any difficulty is encountered, it is worth trying to consolidate your knowledge and skills. Use the information in the student margin to identify the type of question you find most tricky. Be aware that Grade A-type questions are expected to be challenging – these are identified in the answers section.

You will need a pen, a sharp pencil, a clear plastic ruler and a calculator for the best results. A couple of different coloured highlighters could also be handy.

Answers

The expected answers on pages 39–52 and 98–120 give National Standard answers, but, occasionally, there may be other acceptable answers. Each answer has a reference to the level of demand of the question – C for questions at a demand level of Grade C and A for those at a demand level of Grade A. In questions with more than one mark, any C level answers are noted first, for example, CA, CCCA. The answers to the Practice Exams also have commentaries with hints and tips provided alongside. Do not feel you need to use them all!

The commentaries focus on the biology itself as well as providing hints and tips, advice on wording of answers and information about commonly made errors.

Revision

There are 23 Key Areas, so covering two each week would need a 12-week revision programme. Starting during your February holiday should give you time for the exam in May – just! You could use the Revision Calendar at www.hoddergibson.co.uk/ESEP to help you plan and keep a record of your progress.

We wish you the very best of luck!

KEY AREA INDEX GRIDS

Practice Questions

This Key Area index grid will guide you when looking for questions by question type or by Area of Study.

Course Areas		Multiple choice (1 mark)	Structured and extended response (ER) questions					Check
Area of Study	Key Area		Short answer (4 marks)	Data (8 marks)	Experimental (8 marks)	Mini ER (4–6 marks)	Full ER (7–10 marks)	
DNA and the genome	1.1	1–2	1			1		
	1.2	3–4	2					
	1.3	5–6	3			2	1	
	1.4	7–8	4					
	1.5	9–10	5	1	1			102
	1.6	11–12	6					
	1.7	13–14	7			3		
	1.8	15–16	8				2	
	SSI	17–21						☐
Metabolism and survival	2.1	22–23	9				3	
	2.2	24–25	10					
	2.3	26–27	11			4		
	2.4	28–29	12	3	2			95
	2.5	30–31	13					
	2.6	32–33	14			5	4	
	2.7	34–35	15			6		
	SSI	36–39						☐
Sustainability and interdependence	3.1	40–41	16			7	5	
	3.2	42–43	17					
	3.3	44–45	18					
	3.4	46–47	19					
	3.5	48–49	20	2	3	8		103
	3.6	50–51	21					
	3.7	52–53	22				6	
	3.8	54–55	23			9		
	SSI	56–60						☐
Totals		60	92	24	24	45	55	300

Practice Exam A

This Key Area index grid will guide you when looking for questions by question type or by Area of Study.

Course Areas		Paper 1	Paper 2					Check
Area of Study	Key Area	Multiple choice	Short answer	SSI Data	SSI Experimental	Mini extended response	Full extended response	
DNA and the genome	1.1	2, 3						
	1.2	1	5					
	1.3	5	1, 2					
	1.4						18	$\overline{40}$
	1.5	6						
	1.6	4	4					
	1.7	7	3					☐
	1.8	8						
Metabolism and survival	2.1	9, 10, 12						
	2.2	11	9					
	2.3	13	8					
	2.4	14		6				$\overline{40}$
	2.5		7					
	2.6		10					☐
	2.7	15, 16	11					
Sustainability and interdependence	3.1	17, 18, 23, 24	15					
	3.2	21						
	3.3	22, 25	14					
	3.4	19		12		13		$\overline{40}$
	3.5		16					
	3.6		17					
	3.7	20						☐
	3.8							
Total		25	67	9	7	4	8	120

Practice Exam B

This Key Area index grid will guide you when looking for questions by question type or by Area of Study.

Course Areas		Paper 1	Paper 2					Check
Area of Study	Key Area	Multiple Choice	Short Answer	SSI Data	SSI Experimental	Mini extended response	Full extended response	
DNA and the genome	1.1	1, 2		4		5		$\overline{40}$
	1.2		1					
	1.3		2					
	1.4	3, 5	3					
	1.5	4						
	1.6	6						
	1.7	9						☐
	1.8	7	6					
Metabolism and survival	2.1	8, 12			8			$\overline{42}$
	2.2	11	7					
	2.3	15						
	2.4		9					
	2.5	13	12					
	2.6	14	11					☐
	2.7	10	10					
Sustainability and interdependence	3.1	16, 17	14				16	$\overline{38}$
	3.2	18	15					
	3.3	20						
	3.4							
	3.5	22	13					
	3.6	21, 23						☐
	3.7	24						
	3.8	19, 25						
Total		25	68	6	8	4	9	120

Question type: Multiple-choice

≫ HOW TO ANSWER

In your examination, Paper 1 consists entirely of multiple-choice questions. There are 25 questions for 1 mark each. Each question should take about 1.5 minutes and has only **one** correct answer. In practice, some questions might take a bit longer if there is a lot to read or if calculations or genetic crosses are involved. Others can be answered more quickly if they require straightforward recall. The time for these questions is taken up in reading and thinking – there is no writing, only a mark in a grid, although you may need to do some working. You should spend no more than 40 minutes on Paper 1 in your examination.

When tackling multiple-choice questions, read the question thoroughly and try to think of the answer without studying the options. Then look at the options:

▶ If your answer is there, that's the job done.

▶ If you are not certain of an answer, read through the question again and choose the option that seems the best fit.

▶ Or, you can try to eliminate options that you are sure are not correct, before making your choice.

Try not to leave any question without an answer marked – complete the grid for each question as you work through.

For these multiple-choice practice questions, you may circle the letter corresponding to your chosen answer, or write your answers on a separate piece of paper.

Top Tip!

In your examination, any rough working for Paper 1 should be done on the additional space for answers and rough work, provided at the end of the supplied answer booklet.

---

1 The diagram represents the components of a single DNA nucleotide.

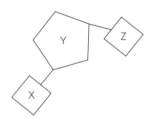

Which row in the table identifies these components?

	X	Y	Z
A	phosphate	sugar	base
B	base	sugar	phosphate
C	sugar	phosphate	base
D	sugar	base	phosphate

Applying KU

2   The following are structures into which DNA can be organised within cells.

   1   linear chromosome

   2   circular chromosome

   3   circular plasmid

Into which of these structures is DNA organised in prokaryotic cells?

   A   1 only

   B   2 only

   C   1 and 2 only

   D   2 and 3 only

Demonstrating KU

3   The statements refer to events in the replication of a lagging strand of DNA.

   1   Primers bind to template chains.

   2   DNA ligase forms sugar–phosphate links.

   3   Hydrogen bonds break.

   4   DNA polymerase adds free nucleotides to strand.

   5   Double helix unwinds.

In which order do these events occur?

   A   1, 2, 4, 3, 5

   B   5, 3, 4, 1, 2

   C   5, 3, 1, 4, 2

   D   1, 3, 2, 4, 5

Demonstrating KU

4   The diagram shows a technique which can be used to separate fragments of DNA to produce a profile.

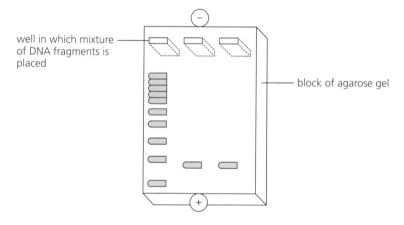

well in which mixture of DNA fragments is placed

block of agarose gel

This technique is

   A   electrophoresis which separates and fragments according to their size and charge

   B   chromatography which separates and fragments according to their size and charge

   C   electrophoresis which separates and fragments according to their solubility in agarose gel

   D   chromatography which separates and fragments according to their solubility in agarose gel.

Applying KU

5 The list shows different ribonucleic acid molecules which occur in living cells.

1 mRNA

2 tRNA

3 rRNA

Which row in the table matches the types of RNA with their functions?

RNA functions		
Picks up and carries specific amino acids	Carries a copy of genetic code	Combines with protein to form ribosomes
A	2	1	3
B	1	3	2
C	2	3	1
D	3	1	2

6 The diagram represents the chemical structure of a folded polypeptide.

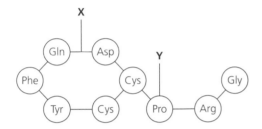

Which row in the table identifies parts X and Y?

	X	Y
A	peptide bond	base
B	peptide bond	amino acid
C	hydrogen bond	base
D	hydrogen bond	amino acid

7 Cellular differentiation occurs because

A cells express some of their genes but not others

B cells all have a different genetic composition

C different cells contain a different set of chromosomes

D different cells lack some genes.

8 There are two types of human stem cell.

1 embryonic stem cells

2 tissue stem cells

Which row in the table describes properties of embryonic and tissue stem cells?

Properties of stem cell		
Self-renewal	Can differentiate	Are multipotent
A	1 only	1 only	both 1 and 2
B	both 1 and 2	both 1 and 2	2 only
C	1 only	both 1 and 2	1 only
D	both 1 and 2	1 only	both 1 and 2

9 Most of a eukaryotic genome consists of

   A   genes that code for proteins

   B   non-coding sequences that code for proteins

   C   genes that code for RNA

   D   non-coding sequences.

Demonstrating KU

10 The list shows three different ribonucleic acid molecules which occur in living cells.

   1   mRNA

   2   tRNA

   3   rRNA

Which row in the table shows features related to the functions of these RNAs?

Demonstrating KU

	RNA features		
	Transcribed from coding regions of DNA	Transcribed from non-coding regions of DNA	Translated to produce polypeptides
A	1 only	2 and 3	1 only
B	2 and 3	1 only	2 only
C	1 only	1, 2 and 3	1 and 2
D	1, 2 and 3	3 only	2 and 3

11 A frame-shift mutation can be the result of

   A   a nucleotide deletion

   B   a gene duplication

   C   a single nucleotide substitution

   D   a nucleotide change at a splice site.

Demonstrating KU

12 A single gene mutation could be a change in the

   A   structure of a chromosome caused by deletion of a gene

   B   number of genes on a chromosome caused by duplication

   C   structure of a chromosome caused by translocation

   D   base sequence of DNA caused by substitution.

Demonstrating KU

13 Bacteria and archaea can exchange genetic material

   A   by both horizontal and vertical transfer between individuals in the same generation

   B   to offspring by horizontal transfer

   C   between individuals in the same generation by vertical transfer only

   D   by both horizontal and vertical transfer.

Demonstrating KU

14 A new species is considered to have evolved when its population

   A   is isolated from other populations by an ecological barrier

   B   shows changes in phenotype frequency due to mutations

   C   can no longer interbreed successfully with other populations

   D   is subjected to increased selection pressure in its habitat.

Demonstrating KU

15 Bioinformatics is the

   A   use of genome information in prescribing drugs

   B   the study of evolutionary history and relationships

   C   use of computers and statistical analysis to compare sequence data

   D   construction of molecular clocks to study evolutionary events.

Demonstrating KU

16 The appearance of the following groups marked key events in the evolution of life on Earth.

1   eukaryotes

2   prokaryotes

3   photosynthetic organisms

4   animals

In which order did these groups appear with the earliest first?

A   2, 3, 1, 4

B   3, 2, 1, 4

C   2, 1, 3, 4

D   3, 4, 2, 1

17 After a few seconds in a thermal cycling PCR machine, a DNA sequence has been amplified to give 64 copies.

How many more cycles of PCR would be needed to amplify the sequence to 2048 copies?

A   4

B   5

C   6

D   11

18 The diagram shows a bacterial cell which has been magnified 600 times.

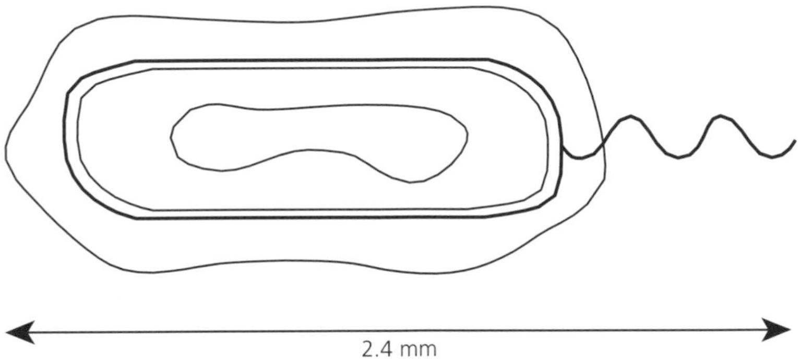

2.4 mm

What is the actual length of the bacterial cell?

A   0.4 μm

B   4.0 μm

C   40 μm

D   0.4 mm

19 Analysis of a molecule of DNA found it to contain 200 cytosine bases, 20% of the total number of bases in the molecule.

How many phosphate groups did it contain?

A   200

B   400

C   800

D   1000

20 The graph shows changes in the number of human stem cells in a culture. The activity of the enzyme glutaminase present in the cells over an eight-day period is also shown.

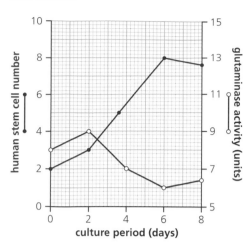

How many units of glutaminase activity were recorded when the cell number was 50% of its maximum over the eight days?

A   8.0

B   8.2

C   9.0

D   13.0

Selecting

21 The sequence of the first 20 amino acids in a protein found in the bones of mammals is shown in the table.

Each amino acid is represented by a capital letter.

Amino acid number																			
Mammal	1				5					10					15				20
human	Y	L	Y	Q	W	L	G	A	P	V	P	Y	P	D	P	L	E	P	R	R
chimpanzee	Y	L	Y	Q	W	L	G	A	P	V	P	Y	P	D	P	L	E	P	R	R
orangutan	Y	L	Y	Q	W	L	G	A	P	V	P	Y	P	D	P	L	E	P	K	R
gorilla	Y	L	Y	Q	W	L	G	A	O	V	P	Y	P	D	P	L	E	P	K	R

Which of the following conclusions is **not** suggested by the data in the table?

A   Humans are most closely related to chimpanzees.

B   Humans and chimpanzees are more closely related to gorillas than to orangutans.

C   These mammals share a close ancestral relationship.

D   Orangutans are more closely related to humans than they are to gorillas.

Concluding

22 The graph shows the energy changes involved in a chemical reaction.

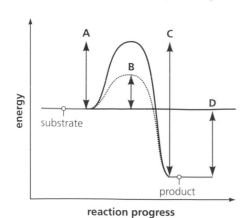

Which letter indicates the activation energy of this reaction in the presence of an enzyme specific to this substrate?

23 The presence of enzyme inhibitors can affect the rates of reaction in a metabolic pathway.

Which row in the table about types of inhibitor is fully correct?

	Type of inhibition	Inhibitor binds to active site?	Effect of increasing substrate concentration on inhibition
A	competitive	no	unaffected
B	non-competitive	no	reversed
C	competitive	yes	unaffected
D	non-competitive	no	unaffected

24 The list shows types of reaction involved in respiration.

1 phosphorylation

2 dehydrogenation

3 fermentation

Which row in the table identifies the types of reaction occurring at the stages in respiration given?

	Stages in respiration	
	Glucose converted to pyruvate in glycolysis	Citrate converted to oxaloacetate
A	2 only	2 and 3
B	1 only	2 only
C	1 and 2	1 and 2
D	3 only	1 and 2

25 The list shows processes in respiration.

X glycolysis

Y fermentation in plant cells

Z fermentation in muscle cells

Which of these processes produces CO_2?

A X only

B Y only

C X and Y only

D X, Y and Z

26 The diagram represents a section through a vertebrate heart showing the direction of blood flow.

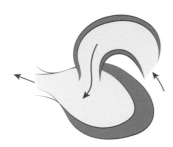

Which row in the table identifies the class of vertebrate with hearts of this type and the type of circulation involved?

	Vertebrate class	Type of circulation
A	fish	double
B	fish	single
C	amphibian	double
D	amphibian	single

27 Which of the following instruments is **not** used to measure metabolic rates?

    A respirometer

    B oxygen probe

    C colorimeter

    D calorimeter

28 Which of the following statements refers to regulators?

    A Their internal environment changes with external environment.

    B They live only within a narrow ecological niche.

    C The maintenance of their internal environment has a high metabolic cost.

    D Their optimum metabolic rate is maintained by behavioural responses alone.

29 The diagram shows an outline of the control of body temperature in a mammal.

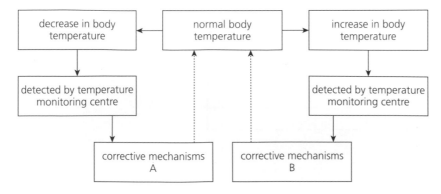

Which are corrective mechanisms at A?

    A vasodilation and contraction of hair erector muscles

    B vasodilation and relaxation of hair erector muscles

    C vasoconstriction and contraction of hair erector muscles

    D vasoconstriction and relaxation of hair erector muscles

30 Garden snails can become dormant in summer when temperatures rise and conditions become dry. The snails respond by entering a state called aestivation in which they seal themselves into their shells and reduce their metabolic rate.

This suggests that aestivation is a

A learned behaviour

B predictive dormancy

C consequential dormancy

D type of hibernation.

Applying KU

31 Young whooping cranes migrate south with their parents or with older birds for their first few years and gradually get better at finding their way to wintering grounds as they age.

This suggests that crane migration behaviour is

A innate

B predictive

C learned

D consequential.

Applying KU

32 Which row in the table matches fermenter conditions with the main reasons for adopting those conditions?

Demonstrating KU

Fermenter conditions		
Sterile at start of lag phase	**Buffers added as required**	**Culture aerated**
A prevents competition	provides optimum conditions for enzymes	supplies oxygen for respiration
B allows enzyme induction	ensures pH stays at 7.0	keeps culture solution well mixed
C prevents competition	ensures pH stays at 7.0	supplies oxygen for respiration
D allows enzyme induction	provides optimum conditions for enzymes	keeps culture solution well mixed

33 The graph shows the growth curve of a population of *Bacillus subtilis* which has been cultured to produce an antibiotic.

Which letter identifies the phase during which genes that encode antibiotics are expressed?

Applying KU

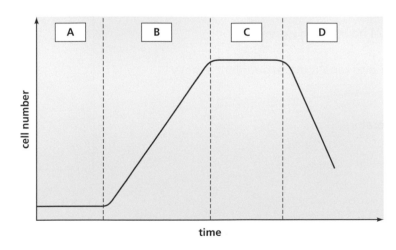

34 The diagram shows a bacterial plasmid that carries genes for resistance to the antibiotics ampicillin and tetracycline.

*Pst*I and *Bam*HI are restriction sites in the plasmid.

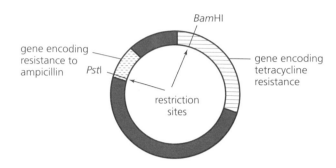

Which row in the table identifies the antibiotics to which bacteria transformed to contain the plasmid would be resistant, if a new gene were inserted into the restriction sites as shown?

	Restriction site into which new gene is inserted	Antibiotic(s) to which transformed bacteria would be resistant
A	*Pst*I	ampicillin
B	*Pst*I	tetracycline and ampicillin
C	*Bam*HI	tetracycline and ampicillin
D	*Bam*HI	ampicillin

Applying KU

35 As part of an experiment into the effect of an antibiotic on the growth of a species of bacteria, an absorbant paper disc soaked in a solution of the antibiotic was placed in the centre of an agar plate.

A pure culture of the bacterial species was then spread over the agar as shown in the diagram.

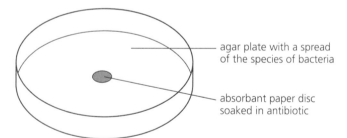

The plate was sealed, incubated at 30°C for 24 hours, then examined for bacterial growth.

A suitable control for this experiment would be to repeat the experiment exactly but

  A   with no bacteria added

  B   incubate the plate at human body temperature

  C   use a disc soaked in water

  D   use a disc soaked in a different antibiotic.

Planning

36 Liver tissue contains an enzyme involved in the breakdown of alcohol. The graph shows the effect of different concentrations of copper ions on the breakdown of alcohol by this enzyme over a 30-minute period.

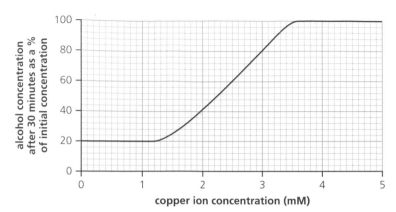

Which of the following conclusions can be drawn from the graph?

A   4.5 mM copper ions has no effect on the enzyme activity.

B   2.5 mM copper ions halves the enzyme activity.

C   0.5 mM copper ions completely inhibits enzyme activity.

D   Enzyme activity increases when copper ion concentration is increased from 1 mM to 2 mM.

Concluding

37 The graph shows how the activity of the enzyme ATPase is affected by the concentrations of sodium and potassium ions.

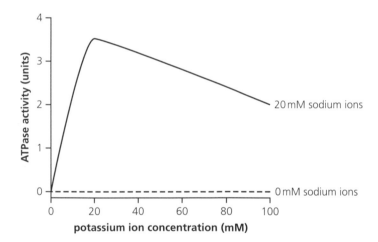

What valid conclusion can be drawn from this information?

A   ATPase activity requires the presence of both sodium and potassium ions.

B   The optimum concentration of sodium ions for ATPase activity is 20 mM.

C   The presence of potassium ions inhibits ATPase activity.

D   ATPase activity requires the presence of sodium ions only.

Concluding

38 The diagram shows the apparatus set up to investigate the rate of respiration in an earthworm.

After 10 minutes at 20°C the level of liquid in the capillary tube had changed as shown.

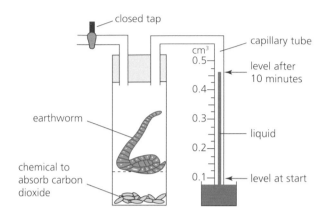

Suggested improvements to the investigation are given below.

1 Use a capillary tube with a narrower diameter.

2 Repeat the experiment several times and take averages.

3 Use a scale with more divisions.

Which of these suggestions would improve the accuracy of the results?

A 1 and 2 only

B 1 and 3 only

C 2 and 3 only

D 1, 2 and 3

Evaluating

39 The graph shows the effect of substrate concentration on the rate of an enzyme-catalysed reaction.

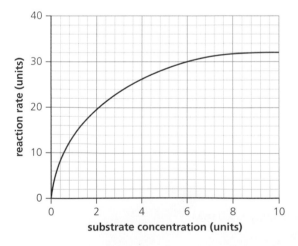

At which substrate concentration is the reaction rate equal to 75% of the maximum rate?

A 2.6

B 3.2

C 6.0

D 7.5

Selecting

40 The following absorption spectra were obtained using a spectroscope and four different plant extracts.

The black areas indicate light absorbed by the extracts.

Which extract contains chlorophyll a?

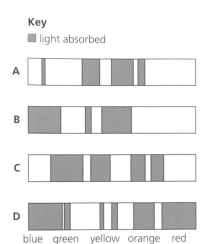

Key
■ light absorbed

A

B

C

D

blue green yellow orange red

41 The transfer of electrons down the electron transport chains in the membranes of chloroplasts takes place as a result of the

A   pigment molecules absorbing light energy

B   photolysis of water

C   ATP synthase generating ATP

D   coenzyme NADP transferring hydrogen ions to the carbon fixation stage.

42 Certain experimental procedures are required when setting up field trials to compare the performance of different cultivars.

Which row in the table matches the procedure with the reason for its inclusion?

| | Reason for inclusion in a field trial | | |
	To ensure valid comparisons	To take account of variability within the sample	To eliminate bias when measuring treatment effects
A	randomisation	selection of treatments	replication
B	replication	randomisation	selection of treatments
C	replication	selection of treatments	randomisation
D	selection of treatments	replication	randomisation

43 Which of the following could explain an increase in the frequency of individuals which are homozygous for recessive deleterious alleles?

A   inbreeding related plants

B   cross breeding animals from different breeds

C   crossing two different inbred lines

D   crossing F_1 hybrids

44 The list describes some properties of weed plants.

1 rapid growth

2 long term seed viability

3 storage organs

4 vegetative propagation

Which of these features are properties of annual weeds?

A 1 only

B 1 and 2 only

C 2 and 4 only

D 3 and 4 only

45 Which of the following is **not** a risk associated with the biological control of pests?

The biological control organism may

A become an invasive species

B prey on other species

C be a pathogen of other species

D bioaccumulate.

46 Which of the following statements could be true of intensive farming methods?

1 It is less ethical than free range farming.

2 It is more labour intensive.

3 It requires more land.

4 It is often more cost effective.

A 1 and 3 only

B 1 and 4 only

C 1, 3 and 4

D 1, 2, 3 and 4

47 Which of the following gives examples of stereotypy and misdirected behaviour as indicators of poor animal welfare?

	Stereotypy	Misdirected behaviour
A	excessive grooming	failure in sexual behaviour
B	apathy	repeatedly biting an object
C	repetitive movements of the whole body	excessive grooming
D	failure in sexual behaviour	hysteria

48 The malaria parasite is taken up from human blood by a mosquito when it feeds. The parasite is able to complete its life cycle in the mosquito, producing infective stages. These are passed on to new hosts when the mosquito takes its next blood meal.

Which row in the table shows the roles of each organism in the life cycle of the malarial parasite?

	Primary host	Secondary host	Vector
A	mosquito	human	mosquito
B	human	mosquito	mosquito
C	mosquito	human	malaria parasite
D	human	mosquito	malaria parasite

49 The greater honeyguide is a bird which feeds on beeswax, bee eggs and grubs, but is unable to open hives to feed. These birds lead humans to beehives they find and wait while the humans open the hives and remove the honey-containing combs. The birds then feed on material from the discarded hives.

Which feeding relationship does this example show?

Applying KU

A mutualism

B predation

C altruism

D parasitism

50 Which of the following statements could be true of cooperative hunting?

Demonstrating KU

1 Less energy is used per individual.

2 May benefit subordinate animals as well as dominant ones.

3 Larger prey may be caught than by hunting alone.

A 1 and 2 only

B 1 and 3 only

C 2 and 3 only

D 1, 2 and 3

51 Some animal species live in social groups for defence.

Which of the following statements describes an advantage of an increase in the size of such a social group?

Applying KU

A Individuals are able to spend less time foraging.

B There are fewer times when more than one individual is looking for predators.

C Each individual can spend more time looking for predators than foraging.

D Individuals are able to spend less time looking for predators.

52 Which of the following best describes an effect of the removal of a dominant heather species on a moorland plant community?

Applying KU

A Lowers species diversity and raises relative abundance of other plant species.

B Raises species diversity and raises relative abundance of other plant species.

C Lowers species diversity and lowers relative abundance of other plant species.

D Raises species diversity and lowers relative abundance of other plant species.

53 The likely effect of the removal of one population of a species of beetle from a oak wood ecosystem is to

Applying KU

A cause extinction of the species

B decrease species diversity in the ecosystem

C create a dominant species

D decrease the genetic diversity of the species.

54 The New Zealand black robin population is slowly recovering from its low point of only five individuals in 1980. Despite every one of the 250 black robins today being a descendant of a single female, called Old Blue, the species is described as viable.

This could be because they

Applying KU

A have a naturally low genetic diversity

B do not seem to be affected by inbreeding problems

C have undergone beneficial mutations

D have naturally low reproductive rates.

55 Asian carp were brought to North America as food, part of the pet trade and for sport fishing. Asian carp are large, feed voraciously and reproduce quickly. They take food and habitat away from native fish and have been known to prey on the eggs of other fish species. This has resulted in some loss of biodiversity.

From the information given, which row in the table best describes the Asian carp?

	introduced	naturalised	invasive
A	✓	X	X
B	✓	X	✓
C	X	X	✓
D	✓	✓	✓

Applying KU

56 The graph shows the effect of applying different levels of fertiliser on the yield of a crop plant.

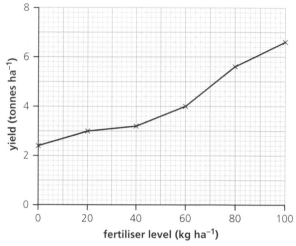

The percentage increase in yield obtained when the fertiliser level is increased from 40 to 80 kg ha^{-1} is

A   24

B   40

C   58

D   75

Selecting & Processing

57 The graph shows changes in the α-amylase concentration and starch content of a barley grain during early growth and development.

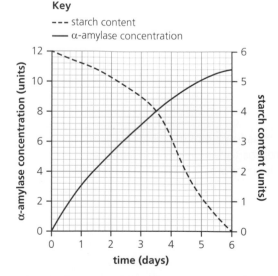

What is the α-amylase concentration when the starch content was 3 units?

A   4.4 units

B   6.0 units

C   8.2 units

D   8.8 units

Selecting

16

58 Two-spotted mites are pests of strawberry plants. An investigation was carried out to provide information on the use of a predatory species of mite to control the two-spotted mites. Predatory mites were released on strawberry plants that were infested with two-spotted mites and the percentage of strawberry leaves occupied by both species was recorded over a 16-week period.

The dependent variable in this investigation was the

A percentage of strawberry leaves occupied by both species

B number of strawberry plants

C time period for the mite interaction

D number of two-spotted mites.

Planning

59 In an investigation into the effects of different wavelengths of light on photosynthesis in green algal cells, apparatus was set up as shown in the diagram. The glass tube containing a suspension of the algal cells was illuminated using filters to provide the different wavelengths of light.

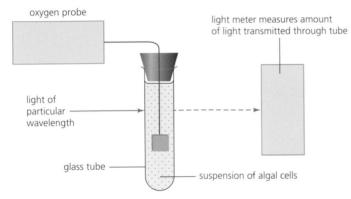

Which of the following is **not** a variable that would need to be kept constant to ensure a valid conclusion could be made?

A colour of filter

B volume of algal suspension

C distance of glass tube from light source

D temperature

Planning

60 The diagram shows the results of a separation of four plant pigments by paper chromatography.

The Rf value of each pigment is calculated by dividing the furthest distance the pigment has moved by the distance the solvent has moved (solvent front) from the origin.

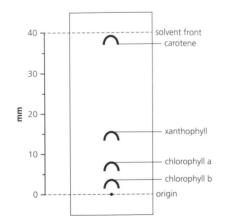

Which pigment has an Rf value closest to 0.4?

A carotene   B xanthophyll

C chlorophyll a   D chlorophyll b

Selecting

Question type: Structured and extended response

Paper 2 of your examination is made up of structured and extended response questions for a total of 95 marks. Again, a mark should take about 1.5 minutes, but questions with lots of reading or thinking time or those with calculations or information to process will take longer, while some straightforward questions can be done more quickly.

In all cases you should pay careful attention to the mark allocation and the number of answer lines or space provided. Each individual mark is awarded separately, so if a question is worth 2 marks there will be two parts to the answer required. If several answer lines are provided, you will probably need to use them.

For the practice questions given here, you should write your answers on a separate piece of paper.

The structured questions are of different types: short answer, data handling and experimental, but these will be mixed together in the actual examination paper and there is some overlap.

There are two types of extended response question: mini extended response and full extended response.

> **Top Tip!**
>
> Remember, Paper 2 of your examination should take no more than 2 hours and 20 minutes.

≫ HOW TO ANSWER

Short-answer questions

Most of the short-answer questions in your exam are focused on testing knowledge. They are often introduced by a short sentence about a Key Area and it is very common to have a labelled diagram presented here. There are likely to be 4–5 marks available for related answers. Very occasionally you may be given a choice of question.

Many of the questions will be at the demand level for Grade C, where you need to name, state or give answers or identify structures. These questions test your memory of the Key Areas (in other words, Demonstrating KU – demonstrating knowledge and understanding).

Some of the questions will be at the demand level for Grade A, which often ask for descriptions, explanations or suggestions. These questions are testing your understanding of Key Areas (that is, Applying KU – application of knowledge and understanding). There are also likely to be some Skills of Scientific Inquiry (SSI) marks mixed in with these questions.

Make sure that you revise your Key Areas thoroughly and systematically, and be aware that the words and terms sought in the answers are those which are given in the Course Specification for Higher Biology – this is crucial.

> **Top Tip!**
>
> - Questions that ask for descriptions, explanations or suggestions often are worth multiple marks, so remember to give a statement for each mark.
> - 'Explain' questions will always require you to bring in correctly selected additional knowledge. There may be several acceptable answers to 'Suggest' questions.

	MARKS	STUDENT MARGIN

1 DNA holds the genetic information in chromosomes in both prokaryotic and eukaryotic cells.

The diagram shows the simplified structure of a prokaryotic and eukaryotic cell.

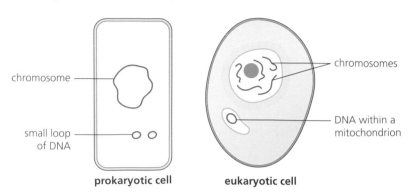

prokaryotic cell   **eukaryotic cell**

a Describe **two** differences between the organisation of DNA in prokaryotic and eukaryotic chromosomes.   **2**   Demonstrating KU

b Name the small loop of DNA shown in the prokaryotic cell.   **1**   Applying KU

c Describe why yeast can be thought of as a special example of a eukaryote.   **1**   Demonstrating KU

2 The diagram shows part of a DNA molecule and other molecules associated with it at a stage in replication in a eukaryotic cell.

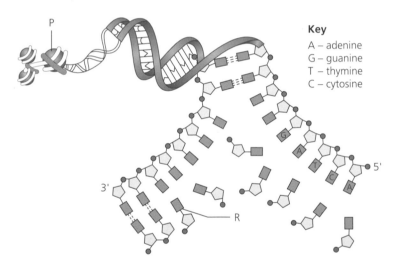

Key
A – adenine
G – guanine
T – thymine
C – cytosine

a Name molecule P, which is associated with the tightly coiled DNA.   **1**   Demonstrating KU

b Name base R.   **1**   Applying KU

c **i** Describe how the diagram illustrates the antiparallel structure of DNA molecules.   **1**   Applying KU

ii The diagram shows synthesis of the leading strand of DNA.

Describe **one** difference between the replication of this strand and the other strand of the molecule.   **1**   Applying KU

MARKS STUDENT MARGIN

3 The diagram shows part of a DNA template strand and part of a primary RNA transcript synthesised from it.

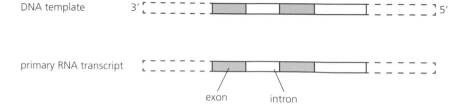

DNA template

primary RNA transcript

exon   intron

a   Name the enzyme involved in the synthesis of a primary transcript. **1**   Demonstrating KU

b   DNA is encoded in triplet sequences.
Explain what is meant by this statement. **1**   Demonstrating KU

c   Name the component of the nucleotide located at the 3′ end of the DNA strand. **1**   Demonstrating KU

d   Use information in the diagram to answer the following question.
Describe how splicing changes the structure of the primary RNA transcript. **1**   Demonstrating KU

4 The diagram shows human embryonic stem cells undergoing differentiation.

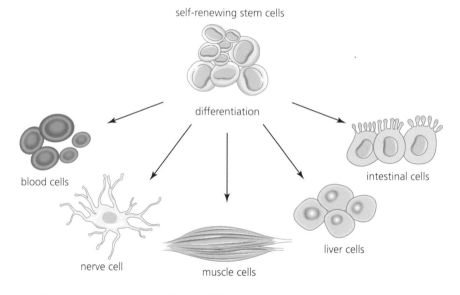

self-renewing stem cells

differentiation

blood cells

nerve cell

muscle cells

liver cells

intestinal cells

a   Explain how a stem cell can differentiate to produce a blood cell. **2**   Applying KU

b   Describe how tissue stem cells differ from embryonic stem cells. **2**   Demonstrating KU

	MARKS	STUDENT MARGIN

5 The diagram shows an image of the linear chromosomes from a cell of a domestic dog.

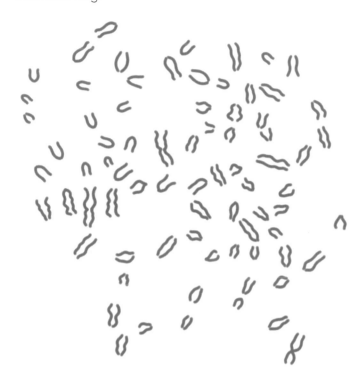

a **i** Explain why these chromosomes do not represent the whole genome of the dog.

1 — Applying KU

ii These chromosomes exist in homologous pairs.

Describe what is meant by a homologous pair of chromosomes.

1 — Demonstrating KU

b Each chromosome has coding and non-coding regions.

i State the difference between a coding and a non-coding region of a chromosome.

1 — Demonstrating KU

ii Give **one** function of non-coding regions of chromosomes.

1 — Demonstrating KU

6 The diagram shows a gene mutation and the effects of this mutation on the mRNA produced when the gene is transcribed.

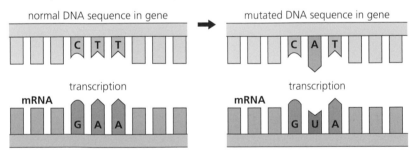

a **i** Name the type of gene mutation which has occurred in this example.

1 — Applying KU

ii Describe the effect this mutation would be expected to have on the protein produced when the mRNA is translated.

1 — Applying KU

b Explain how duplication mutation may have been important in the process of evolution.

2 — Applying KU

	MARKS	STUDENT MARGIN

7 The diagram shows four species of *Anolis* lizard which have evolved on a Caribbean island. The different species have different habitat preferences as shown in the diagram.

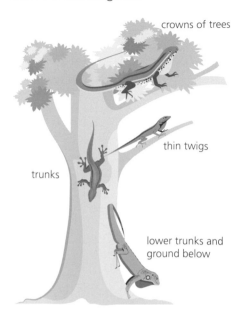

crowns of trees

thin twigs

trunks

lower trunks and ground below

   a   **i** Name the type of isolation which is based on habitat preferences.   **1**   Demonstrating KU

       ii Explain how this type of isolation may have led to the formation of the four *Anolis* species shown.   **2**   Applying KU

   b The *Anolis* lizards are eukaryotic and inherit their genetic material vertically. Evolution can be more rapid in prokaryotes, which can transfer genetic material horizontally.

      Describe how genetic sequences are inherited horizontally.   **1**   Demonstrating KU

8 The phylogenetic tree shows six animal groups in existence today.

Also shown are features which arose at the times shown as the various groups evolved.

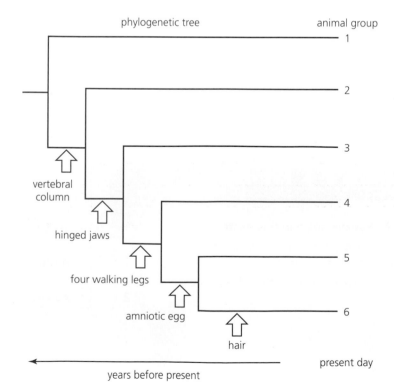

phylogenetic tree   animal group

1

2

3

vertebral column

hinged jaws

4

four walking legs

5

amniotic egg

6

hair

← years before present   present day

		MARKS	STUDENT MARGIN

a Give the number(s) of the phylogenetic group(s) in existence today which do(es) not have hinged jaws. — 1 — Applying KU

b From the information shown, list the features of animal group 3. — 1 — Applying KU

c State **two** sources of evidence which allow accurate phylogenetic trees to be constructed. — 2 — Demonstrating KU

9 The diagram shows four metabolites A–D in a metabolic pathway.

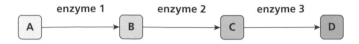

a **i** Describe how the concentration of metabolite D could be controlled by end product inhibition. — 1 — Demonstrating KU

   ii Describe the effect on the concentration of metabolites B and C if a non-competitive inhibitor of enzyme 2 is added. — 2 — Applying KU

b Describe the role of genes in the control of metabolic pathways such as this. — 1 — Applying KU

10 The electron transport chain involved in the final stage of aerobic respiration in cells is made up of proteins in membranes as shown in the diagram.

H⁺ hydrogen ion

e⁻ electron

protein

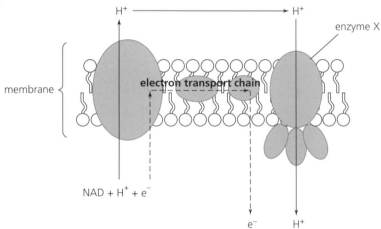

a Describe the **exact** location of these membranes in cells. — 1 — Applying KU

b Describe the role of the electrons in the electron transport chain. — 1 — Demonstrating KU

c Name enzyme X embedded in the membrane. — 1 — Applying KU

d State the role of oxygen in the electron transport chain. — 1 — Demonstrating KU

		MARKS	STUDENT MARGIN

11 Oxygen is consumed during aerobic respiration and organisms such as birds that have high metabolic rates require the efficient delivery of oxygen to their cells.

   a Describe **two** methods of measuring the metabolic rate of an organism. `2` *Demonstrating KU*

   b The diagram shows a section through the heart of a bird.

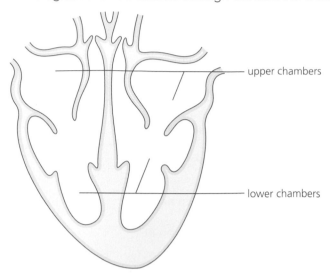

   upper chambers

   lower chambers

   With reference to the heart chambers, explain how the structure of the bird's heart makes it more efficient at delivering oxygen to muscles than the heart of a fish. `2` *Applying KU*

12 The ability of an organism to maintain its metabolic rate is affected by external abiotic factors.

   a Name **one** external abiotic factor, other than temperature, which can affect the ability of an organism to maintain its metabolic rate. `1` *Demonstrating KU*

   b Give the term used to describe organisms with the ability to control their internal environment by metabolic means. `1` *Demonstrating KU*

   c Give **two** reasons why body temperature in humans is important to metabolic processes. `2` *Demonstrating KU*

13 The diagram represents the movements of wildebeest in East Africa over a period of a year.

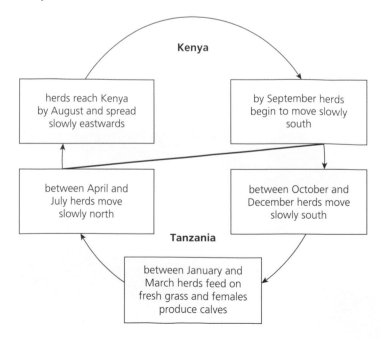

Kenya

herds reach Kenya by August and spread slowly eastwards

by September herds begin to move slowly south

between April and July herds move slowly north

between October and December herds move slowly south

Tanzania

between January and March herds feed on fresh grass and females produce calves

			MARKS	STUDENT MARGIN

a **i** Name the behavior which involves the relocation of the wildebeest through the year.

1 Applying KU

ii Using information in the diagram, suggest **one** advantage and **one** disadvantage to the wildebeest of relocating.

2 Applying KU

b Wildebeest are large animals which move slowly over land and scientists are able to track them by following the herds in vehicles and using photography to record their progress.

Describe how scientists have been able to track small fast-flying animals such as birds, bats and some insects.

1 Demonstrating KU

14 The graph shows the growth phases in a culture of microorganisms in a fermenter.

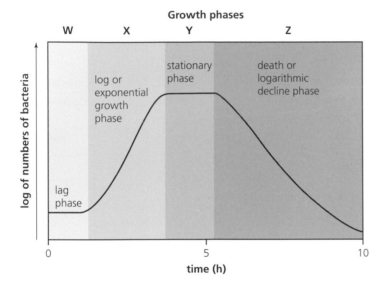

Growth phases

a Use letters from the graph to identify the growth phases in which the following events occur.

i Secondary metabolites are synthesised.

1 Applying KU

ii Enzymes are induced.

1 Applying KU

b Explain the importance to the microorganisms of:

i the synthesis of secondary metabolites

1 Demonstrating KU

ii inducing enzymes.

1 Demonstrating KU

15 Wild strains of microorganisms with the potential to be used in industry can be improved by mutagenesis and recombinant DNA technology.

a **i** State what is meant by the term mutagenesis.

1 Demonstrating KU

ii Explain why mutagenesis is **not** a reliable way to produce new strains of microorganisms.

1 Demonstrating KU

b In recombinant DNA technology, modified plasmids are sometimes added to bacteria to produce the desired new strain.

Give **two** features of a bacterial plasmid used in recombinant DNA technology which allow it to be an effective vector.

2 Demonstrating KU

MARKS   STUDENT MARGIN

16 The diagram shows some events in the carbon fixation stage (Calvin cycle) of photosynthesis in green plant cells kept illuminated in bright light.

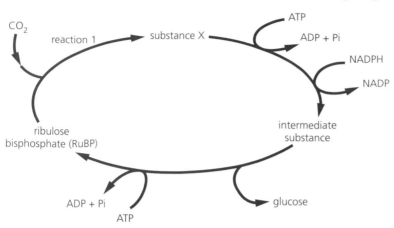

a Name **both** the enzyme which catalyses reaction 1 **and** substance X.

2   Applying KU

b Describe **one** fate of the glucose produced by photosynthesis.

1   Demonstrating KU

c Describe the change in concentration of RuBP in an illuminated green plant cell if its source of carbon dioxide is removed.

1   Applying KU

17 Pure bred cattle can be affected by inbreeding depression and so may have poorer health and vigour characteristics than crossbreeds. The chart shows the average milk yield from two pure breeds of cattle and their F_1 crossbreed offspring.

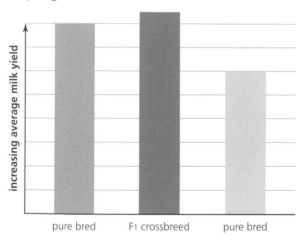

a **i** Describe what the chart shows about the effect of crossbreeding on milk yield.

1   Applying KU

ii The F_1 offspring are heterozygous for many of their characteristics.

Explain what is meant by the term heterozygous.

1   Demonstrating KU

iii Explain why the F_1 crossbreed cattle are not bred together to produce F_2 animals for milk.

1   Demonstrating KU

b Recombinant DNA technology has allowed the introduction of Bt toxin genes into certain crop plant varieties.

Explain the advantage of this in terms of food production.

1   Demonstrating KU

MARKS   STUDENT MARGIN

18 Buttercups are perennial weeds that can grow among crop plants where they reduce yields and productivity.

    a   Describe what is meant by a perennial weed **and** give **one** adaptation which allows them to be successful weeds of crop plants.

2   Demonstrating KU

    b   Systemic herbicides such as glyphosate are weed-killers which can be used against perennial weeds. In recent years, glyphosate-resistant crop varieties have been produced using GM techniques.

       i   Describe the advantage of using systemic herbicides against perennial weeds.

1   Demonstrating KU

       ii   Explain the advantage of developing glyphosate-resistant crop varieties.

1   Applying KU

19 Intensive farming can sometimes create conditions of poor animal welfare. This is often indicated by changes in the animals' behaviour.

    a   Name **two** examples of behaviour which could indicate poor animal welfare.

2   Demonstrating KU

    b   Intensive farming is less ethical than free-range farming due to poorer animal welfare.

       Give **one** example of a cost and **one** example of a benefit associated with the adoption of free-range farming methods.

2   Demonstrating KU

20 a   The diagram shows the life cycle of a parasitic worm which causes schistosomiasis in humans.

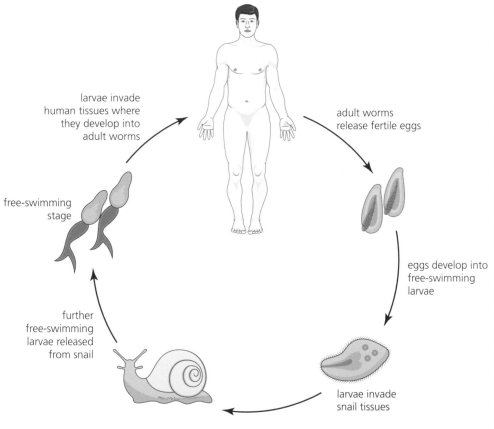

larvae invade human tissues where they develop into adult worms

adult worms release fertile eggs

free-swimming stage

eggs develop into free-swimming larvae

further free-swimming larvae released from snail

larvae invade snail tissues

       i   Describe the features of the parasite's relationship with its host.

1   Demonstrating KU

       ii   Explain why a parasite cannot survive for long out of contact with its host.

1   Demonstrating KU

       iii   The snail is an intermediate host for this parasite.

          Give evidence that the snail is not acting as a vector in this example.

1   Applying KU

    b   Describe what is meant by the term mutualism.

1   Demonstrating KU

		MARKS	STUDENT MARGIN

21 a Long-tailed tits *Aegithalos caudatus* are small birds which breed in loose colonies in which many of the birds are related.

If a pair fails in breeding, they will often behave in an altruistic way by assisting neighbouring pairs to feed their chicks.

Give the meaning of the term 'altruistic' **and**, from the information given, suggest an explanation for the birds' behaviour. **2** Applying KU

b Domestic honey bees *Apis mellifera* are social insects.

Describe the reproductive contribution made by drones **and** workers in a honey bee colony. **2** Demonstrating KU

22 Heather moor ecosystems in Scotland are often dominated by the species *Calluna vulgaris*. The presence of this dominant species has an effect on the species diversity of the community.

a Describe **two** aspects of species diversity. **2** Demonstrating KU

b Describe the effect of the presence of a dominant species such as *Calluna vulgaris* on the species diversity of its community **and** suggest a reason for this effect. **2** Demonstrating KU

23 The signal crayfish, *Pacifastacus leniusculus*, is a North American species introduced to Europe in the 1970s to supplement native crayfish fisheries. It has increased its population and spread rapidly in its new habitats and is now considered an invasive species across Europe.

a Suggest why the populations of signal crayfish have become very large in its new habitats. **1** Applying KU

b Give **one** reason why large populations of invasive species such as signal crayfish can reduce the biodiversity of an ecosystem in which they have become established. **1** Demonstrating KU

c Plans are being made to control the signal crayfish population in Scotland.

Suggest a method of biological control by which the crayfish numbers might be reduced **and** give a possible drawback of this method. **2** Applying KU

›› HOW TO ANSWER

Data questions

These questions focus on the results of an experiment or investigation which has been carried out. There will be some results data given as either a table or a graph. You will have to answer about 8 marks worth of questions. You might be asked to draw a graph or complete a pie chart if the results are presented as a table.

Top Tip!

Draw graphs and bar charts using a ruler and use the data table headings and units for the axis labels. Marks are given for providing scales, labelling the axes correctly and plotting the data points/drawing the bars. Line graphs require points to be joined with straight lines using a ruler.

Top Tip!

If you are given space for calculation, you will very likely need to use it!

You may have to read the data and select specific values. There will very often be calculations from the values – average, ratio or percentage change are the most common, but you can be asked to do simple + − × ÷ sums as well.

You might be asked to predict results for situations not tested – for example, higher or lower temperature, pH or light intensity. Predictions can involve extending graph trends or reading between values in a table.

Make sure you have a sharp pencil, a ruler and a calculator to help with these questions. A highlighter pen can also help.

Top Tip!

Remember to use values from the graph when describing graph trends if you are asked to do so.

If you are asked to calculate an increase or decrease between points on a graph, you should use a ruler to help accuracy – draw pencil lines on the actual graph if this helps.

	MARKS	STUDENT MARGIN

1 The **phylogenetic tree** shows the divergences of some modern animal groups. The **molecular clock** shows rate of change in the percentage of differences in the amino acid sequence in the protein cytochrome C found in these animal groups which occurred over the period of time covered by the phylogenetic tree.

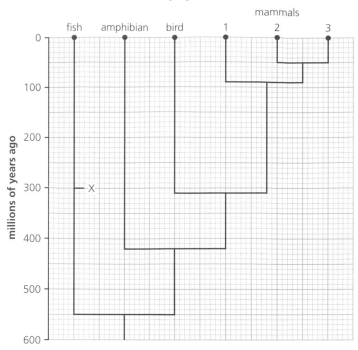

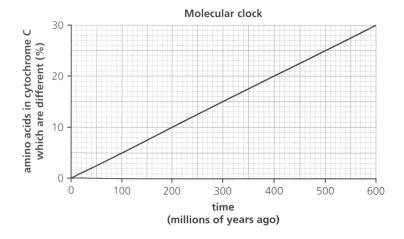

a From the **phylogenetic tree**:

i identify the time when the last common ancestor of the mammals and birds existed — **1** — Selecting

ii calculate the difference in time between the divergence of mammals 1 and 3 and the divergence of mammals 2 and 3 — **1** — Processing

iii predict the age of a fossil bone which was found to be from a common ancestor of mammals 2 and 3. — **1** — Predicting

b From the **molecular clock**:

i calculate the average rate of change per million years in the amino acid sequence of cytochrome C — **1** — Processing

ii predict how long ago an animal with cytochrome C containing 35% of amino acids different to those in modern cytochrome C existed. — **1** — Predicting

c Amino acid sequences in the cytochrome C in two vertebrates differed by 25%.

Predict how long ago their last common ancestor existed. — **1** — Predicting

d From the **phylogenetic tree** and the **molecular clock**:

    i give the percentage difference in the amino acid sequence of cytochrome C in birds compared with amphibians

        1   Selecting

    ii give the percentage difference between the amino acid sequence of cytochrome C which would be expected between a modern fish and an ancestral species found at X on the phylogenetic tree.

        1   Selecting

2 Canola is a plant grown as a crop because its seeds are rich in oil. The extracted oil is used in cooking and as a sustainable biofuel.

Scientists carried out an investigation into the effect of nitrate fertiliser on the seed yield and percentage of oil in the seeds of Canola.

The graph shows the results of this investigation.

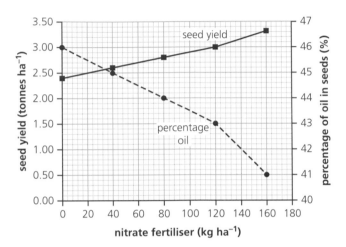

a Use values to describe the relationship between the nitrate fertiliser application, seed yield and percentage of oil in seeds.

        2   Selecting

b Using information in the graph, calculate the percentage change in seed yield when the level of nitrate fertiliser is increased from 0 to 160 kg ha⁻¹.

        1   Processing

c Calculate the average decrease in percentage oil per kg of fertiliser between 0 and 80 kg ha⁻¹.

        1   Processing

d Give the percentage of oil in seeds when the seed yield was 2.8 tonnes ha⁻¹.

        1   Selecting

e Express, as the simplest whole number ratio, the seed yield to kg of fertiliser per hectare, when the percentage of oil is 43%.

        1   Processing

f Predict the percentage of oil content in seeds if a nitrate application of 180 kg ha⁻¹ was given.

        1   Predicting

g Suggest how the validity of the investigation could have been improved.

        1   Evaluating

3 Ruby-throated hummingbirds *Archilochus colubris* have very high metabolic rates during normal activity. They migrate thousands of kilometres each year between their wintering areas in Central America and their summer breeding areas in parts of North America.

They feed on nectar throughout the year and save energy in cold conditions during the night by entering a temporary state known as torpor during which their metabolic rates are much reduced.

MARKS   STUDENT MARGIN

Graph 1 shows how the average body masses of the birds changed over a yearly cycle and **Graph 2** shows how the birds' average metabolic rate during normal activity and torpor changed with air temperature.

Graph 1

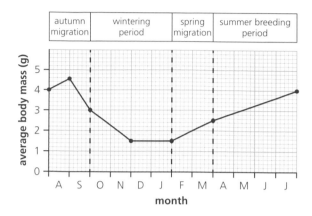

Graph 2

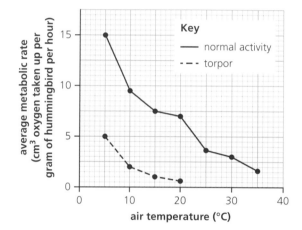

a   i   Use values from **Graph 1** to describe the changes in average body mass of the hummingbirds from the beginning of August to the end of November.

2   Selecting

ii   Calculate the percentage increase in average body mass during the summer breeding period.

1   Processing

iii   Suggest **one** reason for the increase in body mass of the birds during summer.

1   Concluding

b   i   Calculate the average decrease in oxygen consumption per gram per hour for each degree reduction in air temperature from 5 °C to 35 °C during normal activity.

1   Processing

ii   Calculate the reduction in metabolic rate of a bird which enters torpor from normal activity at 10 °C.

1   Processing

iii   Suggest why no values are given for metabolic rate during torpor above 20 °C.

1   Concluding

c   Using information from **Graphs 1** and **2**, calculate the average volume of oxygen consumed per hour by a hummingbird, at the end of September, during normal activity at 20 °C.

1   Processing

≫ HOW TO ANSWER

Experimental questions

These questions focus on an experiment or investigation which has been carried out. Often there is a very short description of the experimental method, sometimes with a diagram of the apparatus used, and some results data given as a table or a graph.

Top Tip!

Try to bring all the information together in your mind to visualise what has been done – doing a little pencil drawing in the margin can sometimes help.

Top Tip!

You should be able to identify the independent and dependent variables in an investigation or experiment.

- The independent variable is the input variable – it usually appears in the first column of a data table and is plotted on the x-axis of a graph.
- The dependent variable refers to the data (results) produced – it usually appears in the second column of a data table and is plotted on the y-axis of a graph.

You will have to answer about 8 marks worth of questions. These can ask for evaluation of, or comments about, planning – such as controls, validity or reliability. Any variable that could affect the results of an experiment should be controlled. A control experiment allows a comparison to be made and allows you to relate the dependent variable to the independent one. The control should be identical to the original experiment apart from the one factor being investigated.

You might be asked to draw a graph if the results are presented as a table. The questions may involve calculations from the results – average, ratio or percentage change are the most common. You might also be asked to predict results for situations not tested – for example, higher or lower temperature, pH or light intensity. Predictions can involve extending graph trends or reading between values in a table.

You may also be asked to draw valid conclusions, giving explanations supported by evidence.

Make sure you have a sharp pencil, a ruler and a calculator to help with these questions. A highlighter pen can also help.

Top Tip!

When concluding, you must refer to the experimental aim, which is likely to be stated in the stem of the question.

		MARKS	STUDENT MARGIN

1 Patients requiring an organ transplant are tissue typed to match with potential donors. Polymerase chain reaction (PCR) is used to amplify DNA from a patient and from potential donors. Gel electrophoresis is used to compare DNA sequences of the patient with those of donors. The presence of specific DNA bands in the gel can indicate that a donor is a suitable match.

In a procedure, amplified DNA from a patient and three potential donors was compared by electrophoresis as shown in the diagram. The table shows how the size of a DNA fragment affects the distance it travels in the gel.

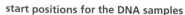

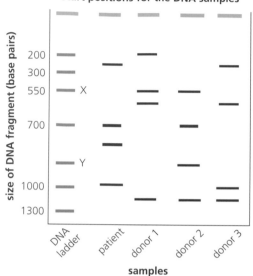

start positions for the DNA samples

Size of DNA fragment (base pairs)	Distance travelled (mm)
200	72
300	58
550	32
700	18
800	12
1000	10
1300	8

a Identify **two** variables which must be controlled when carrying out electrophoresis to compare two DNA samples. — 2 — Planning

b Suggest what should be done to increase the reliability of the procedure. — 1 — Evaluating

c Identify the distance travelled by fragment X. — 1 — Selecting

d **On a piece of graph paper**, draw a line graph to show that the size of a DNA fragment affects how far it travels in the gel. — 1 — Presenting

e Describe the relationship between the size of a DNA fragment and the distance it travels in the gel. — 1 — Selecting

f Fragment Y has travelled 900 mm in the gel.

Give the number of base pairs in this fragment. — 1 — Processing

g Identify the donor whose DNA best matches that of the patient. — 1 — Concluding

	MARKS	STUDENT MARGIN

2   The apparatus shown in the diagram was used to measure the rate of respiration of germinating seeds in air. The distance moved by the drop of coloured liquid was measured at 15-minute intervals for one hour.

The experiment was repeated using woodlice instead of seeds.

The table shows the results.

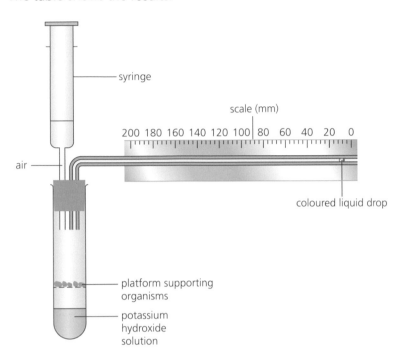

Organism	Distance moved by liquid in each 15-minute interval (mm)				Mean rate of respiration (mm/min)
Germinating seeds	8	7	9	6	0.5
Woodlice	14	13	13	14	

a   Calculate the mean rate of respiration for the woodlice, expressed as movement of coloured liquid drop in millimetres per minute.

b   i   Explain how the respiration of the organisms and the presence of the solution in the apparatus account for the movement of the coloured liquid on the scale.

ii   Suggest **one** precaution that would be required to allow a valid comparison between the mean rates of respiration of the germinating seeds and the woodlice.

iii   Control apparatus should be included in this investigation.

Describe the control **and** explain its purpose in the investigation.

iv   Suggest how the temperature of the tubes could be kept constant.

v   Suggest the purpose of the syringe in this apparatus.

MARKS	STUDENT MARGIN
1	Processing
2	Concluding
1	Evaluating
2	Planning
1	Planning
1	Planning

MARKS   STUDENT MARGIN

3 An experiment on the effect of light on the Calvin cycle was carried out using *Chlorella*, a unicellular alga. 500 cm³ of *Chlorella* was placed in a glass flask in illuminated conditions for 10 minutes and allowed to photosynthesise normally, as shown in the diagram.

After this time, samples were withdrawn from the flask into the collecting vessel every minute for 5 minutes.

The lamp was then switched off and a further set of samples taken every minute for 5 minutes.

The concentrations of ribulose bisphosphate (RuBP) and 3-phosphoglycerate (3PG) in each sample were determined and the results are shown in the table.

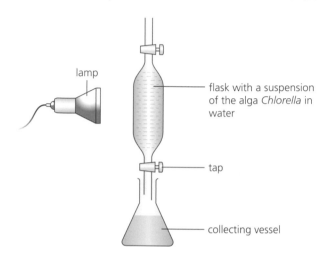

lamp

flask with a suspension of the alga *Chlorella* in water

tap

collecting vessel

Time (minutes)	Illumination	RuBP concentration (units)	3PG concentration (units)
1	light	8	20
2	light	8	20
3	light	8	20
4	light	8	20
5	light	8	20
6	dark	11	18
7	dark	14	16
8	dark	17	14
9	dark	20	12
10	dark	20	12

a **i** Identify the independent variable in this investigation.

1   Planning

ii Identify **one** variable which should have been kept constant within the flask during the experiment to ensure that the procedure was valid.

1   Planning

b Give **one** reason why the flask was left for 10 minutes before the sampling started.

1   Planning

c Describe how the experimental procedure could be improved to increase the reliability of the results.

1   Evaluating

d **On a piece of graph paper**, draw a line graph to show the concentration of RuBP against time.

2   Presenting

e Using the results from the table, describe the evidence which suggests that RuBP is converted to 3PG during the Calvin cycle of photosynthesis.

2   Concluding

>> HOW TO ANSWER

Mini extended response questions

There will usually be a maximum of two open-ended questions in your Paper 2 examination, for which you will need to give extended responses. There could be a choice of question, but not always.

Each question will be short, but several answer lines will be given, which will be a good clue to the answer length. You will need several sentences for a full answer in each case, and there will be 4–6 marks available for each question. Each mark is awarded separately, so the mark allocation gives a clue to the expected answer length too.

The questions test your understanding of related knowledge so you could be asked to describe a process or to compare structures or processes. If you are asked to describe a process, remember to be logical, starting from the beginning of the process and working through in steps. If you are asked to compare two processes or structures, ensure that you describe both of them in full.

Top Tip!

Read the question very carefully. If there is a choice, be clear about which you are selecting.

Top Tip!

'Give an account of' means the same as 'describe'.

		MARKS	STUDENT MARGIN
1	Give an account of the process of cellular differentiation in stem cells.	5	Demonstrating KU
2	Give an account of the structure of the genome.	5	Demonstrating KU
3	Give an account of phylogenetics and molecular clocks.	5	Demonstrating KU
4	Give an account of the competitive and non-competitive inhibition of enzymes.	5	Demonstrating KU
5	Describe and compare metabolism in conformers and regulators.	5	Demonstrating KU
6	Describe the features of a plasmid required for it to function as an effective vector in recombinant DNA technology.	5	Demonstrating KU
7	Give an account of food security and sustainability.	5	Demonstrating KU
8	Give an account of plant field trials.	5	Demonstrating KU
9	Write notes on components of biodiversity and how they are measured.	5	Demonstrating KU

≫ HOW TO ANSWER

Full extended response questions

There will be one full extended response question at the end of Paper 2, which always contains a choice.

Some extended response questions are divided into two or even three parts. It's best to answer each part separately under its heading. There are about 7–10 marks for this question and you will need to make a statement for each mark.

Like the mini extended response questions, these questions test your understanding of related knowledge, so you could be asked to describe a process or to compare structures or processes. If you are asked to describe a process, remember to be logical, starting from the beginning of the process and working through in steps. You are allowed to include diagrams in your answer, but ensure that these are labelled.

Top Tip!

Read the questions very carefully and spend a minute or two to decide which choice is best for you, in other words, which one you can recall more about.

Top Tip!

Each mark is awarded for a separate statement, so the mark allocation gives a clue to the expected answer length.

		MARKS	STUDENT MARGIN
1	Give an account of DNA molecules under the following headings:		
	a their organisation in living cells	5	Demonstrating KU
	b their amplification by the polymerase chain reaction (PCR).	4	
2	Describe the evolution of new species under the following headings:		
	a isolation and mutation	5	Demonstrating KU
	b natural selection.	4	
3	Give an account of the electron transport chain and the transfer of energy by ATP.	8	Demonstrating KU
4	Give an account of the production of protein by recombinant DNA technology.	10	Demonstrating KU
5	Give an account of chemical methods used to protect crop plants and the environmental damage which may result from their use.	9	Demonstrating KU
6	Write notes on social behaviour in animals under the following headings:		
	a social hierarchy and cooperative hunting	6	Demonstrating KU
	b social insects.	4	

Multiple-choice

Question	Answer	Mark	Demand
1	B	1	A
2	D	1	C
3	C	1	A
4	A	1	C
5	A	1	C
6	B	1	C
7	A	1	C
8	B	1	A
9	D	1	A
10	A	1	C
11	A	1	C
12	D	1	C
13	D	1	A
14	C	1	C
15	C	1	C
16	A	1	C
17	B	1	C
18	B	1	C
19	D	1	A
20	B	1	A
21	C	1	C
22	B	1	C
23	D	1	A
24	C	1	A
25	B	1	C
26	B	1	C
27	C	1	C
28	C	1	A
29	C	1	C
30	C	1	C
31	C	1	C
32	A	1	A
33	C	1	A
34	D	1	C
35	C	1	A
36	B	1	C
37	A	1	A
38	C	1	C
39	B	1	C
40	D	1	C
41	A	1	C
42	D	1	A

Question	Answer	Mark	Demand
43	A	1	A
44	B	1	C
45	D	1	A
46	B	1	C
47	C	1	A
48	B	1	A
49	A	1	A
50	D	1	A
51	D	1	A
52	B	1	A
53	D	1	A
54	A	1	A
55	B	1	A
56	D	1	A
57	D	1	C
58	A	1	A
59	A	1	C
60	B	1	C

Short-answer questions

Question			Expected answer	Mark	Demand
1	a		Linear chromosomes in eukaryotes **AND** circular chromosome in prokaryotes **OR** Eukaryotic chromosomes located in the nucleus but prokaryotes have no nucleus **OR** Eukaryotic chromosomes are associated/tightly coiled with histones **OR** Eukaryotes contain circular chromosomes in mitochondria/chloroplasts/other organelles **(Any 2)**	2	CA
	b		Plasmid	1	C
	c		Contain plasmids	1	C
2	a		Histone	1	C
	b		Thymine	1	C
	c	i	The 3′ (deoxyribose) end of one strand is bonded/joined to the 5′ (phosphate) end of its complementary strand	1	A
		ii	The leading strand is replicated continuously **AND** the lagging strand/other strand is replicated/built up in sections/fragments	1	A
3	a		RNA polymerase	1	C
	b		A sequence of 3 bases codes for a specific amino acid	1	A
	c		Deoxyribose	1	A
	d		Removes introns **AND** joins exons together	1	A

Question			Expected answer	Mark	Demand
4	a		Genes encoding proteins characteristic of red blood cells are expressed/switched on = **1** Genes encoding other proteins are not expressed/switched off = **1**	2	CA
	b		Tissue cells can differentiate into any cell from their tissue/are multipotent = **1** Embryonic stem cells can differentiate into (nearly) any cell type in the organism/are pluripotent = **1**	2	CA
5	a	i	There are other DNA sequences in the mitochondria	1	A
		ii	They contain the same genes in the same order	1	A
	b	i	Coding regions encode protein/are genes **AND** non-coding regions do not encode proteins/are not genes	1	A
		ii	Non-coding sequences are regulatory sequences/regulate transcription/turn genes on or off **OR** Non-coding sequences are transcribed but not translated/transcribed into tRNA/transcribed into rRNA	1	A
6	a	i	Substitution	1	C
		ii	One amino acid will be different	1	C
	b		A second copy of a gene is present = **1** A single gene mutation in a duplicated region of a chromosome can produce an advantageous gene/favourable characteristic without the loss of an existing gene = **1**	2	CA
7	a	i	Sympatric	1	C
		ii	Sub-population remains within own preferred habitat = **1** Do not breed with/exchange genes with other populations = **1**	2	CA
	b		Plasmids/chromosomes passed from one member of a generation to another member of the same generation	1	C
8	a		1 **AND** 2	1	C
	b		Vertebral column **AND** hinged jaws = **1**	1	CA
	c		Fossil record = **1** Sequence data = **1**	2	CA
9	a	i	Build up of D acts as an inhibitor of an earlier step in the pathway so leads to reduced concentration of D	1	A
		ii	B will increase = **1** C will decrease = **1**	2	CC
	b		Genes encode the enzymes which control the pathway	1	A
10	a		Inner membrane of the mitochondria	1	C
	b		Release/provide energy to pump the hydrogen ions across the inner membrane of the mitochondria	1	A
	c		ATP synthase	1	C
	d		Acts as the final hydrogen ion **AND** electron acceptor	1	A
11	a		Oxygen probe can measure the oxygen consumed in a given period of time = **1** Carbon dioxide probe can measure the gas produced in a given period of time = **1** A calorimeter can measure the energy released as heat in a given period of time = **1** (Any 2)	2	CA

	Question		Expected answer	Mark	Demand
	b		Upper chambers receive blood from lungs and body at the same time so blood kept separate **OR** No mixing of oxygenated and deoxygenated blood = **1** Lower chambers pump blood to lungs and body at the same time so blood pressure maintained **OR** Blood pumped at high pressure to the muscles = **1**	2	CA
12	a		Salinity/pH	1	C
	b		Regulator	1	C
	c		Enzymes have an optimum temperature at which they are most active = **1** Diffusion rates are affected by temperature = **1**	2	CA
13	a	i	Migration	1	C
		ii	Advantage: avoid metabolic adversity/adverse conditions = **1** Disadvantage: use up energy in movement = **1**	2	CA
	b		Ringing **AND** recapture **OR** Tags using GPS **OR** Transmitters **AND** receivers/satellites **(Any 1)**	1	C
14	a	i	Y	1	C
		ii	W	1	C
	b	i	Gives organisms a competitive/ecological advantage	1	A
		ii	Saves energy/resources **OR** Enzymes only synthesised when needed **OR** Allows use of available substrate	1	A
15	a	i	The process of inducing mutations	1	C
		ii	Mutation is random	1	A
	b		Restriction site **OR** Origin of replication **OR** Selectable marker gene **OR** Regulatory sequence **OR** (Safety) genes preventing survival in external environment **(Any 2)**	2	CA
16	a		Enzyme: RuBisCo = **1** Substance X: 3PG/3 phosphoglycerate = **1**	2	CC

Question			Expected answer	Mark	Demand
	b		Used in respiration/as a respiratory substrate **OR** Stored as starch **OR** Synthesised/converted to cellulose **OR** Used in the synthesis of a variety of metabolites/DNA/proteins/fat **(Any 1)**	1	C
	c		Increase	1	A
17	a	i	The average milk yield of the F_1 crossbreed population is greater than that of both the pure breed populations	1	C
		ii	Have two different alleles of the same gene	1	A
		iii	The F_2 population/individuals/generation would be genetically variable/have a wide variety of genotypes and may have a reduced/lower milk yield	1	A
	b		Insertion of the Bt toxin gene into plants gives them pest resistance and so increases yields	1	C
18	a		Grows back year after year = **1** Storage organs **OR** vegetative reproduction = **1**	2	CA
	b	i	Kill all parts of the weed so it cannot ever grow back	1	C
		ii	Whole field can be sprayed and crops survive while weeds are killed	1	C
19	a		Stereotypy **OR** Misdirected behaviour **OR** Failure in sexual or parental behaviour **OR** Altered levels of activity/very low levels of activity (apathy)/very high levels of activity (hysteria) **(Any 2)**	2	CC
	b		Cost: free range requires more land **OR** is more labour intensive = **1** Benefit: produce can be sold at a higher price **OR** animals have a better quality of life = **1**	2	CA
20	a	i	Parasite benefits in terms of energy/nutrients **AND** host is harmed by the loss of these resources	1	A
		ii	Has limited metabolism	1	C
		iii	The transfer to host is by free-swimming larvae/stages **OR** the snail does not actively transmit the parasite	1	A
	b		Both species in a co-evolved/intimate association benefit	1	C
21	a		Meaning: a behaviour that harms the donor individual but benefits the recipient = **1** Explanation: donor/birds benefit in terms of increased chances of survival of shared genes in the recipients/related birds' offspring **OR** their own future offspring will similarly benefit by reciprocal behaviour = **1**	2	CA
	b		Drones: male sex in reproduction **OR** fertilise queen's eggs = **1** Workers: cooperate to raise offspring **OR** collect food/pollen/nectar **OR** clean/defend hive = **1**	2	CA

	Question		Expected answer	Mark	Demand
22	a		Number of different species/species richness = **1**	2	CA
			Relative abundance of each species = **1**		
	b		Reduces the species diversity = **1**	2	CA
			Takes majority of light/water/resources **OR** outcompetes other species = **1**		
23	a		May be free of predators/parasites/pathogens/ competitors which occur in their native habitat	1	C
	b		May prey on native species **OR** may outcompete native species for resources **OR** may hybridise with native species **OR** may lead to habitat degradation **OR** may lead to increased grazing pressures	1	C
	c		Method: introduce natural predator = **1**	2	CA
			Drawback: introduced predator may become invasive/affect other native organisms/disrupt food web/reduce biodiversity = **1**		
			OR		
			Method: introduce pathogen/disease = **1**		
			Drawback: disease/(microbial) pathogen may affect native/non-target species = **1**		

Data questions

	Question		Expected answer	Mark	Demand
1	a	i	310 million years ago	1	C
		ii	40 million years	1	C
		iii	Any value within the range 50–90 million years ago	1	C
	b	i	0.05% per million years	1	A
		ii	700 million years	1	A
	c		500 million years	1	A
	d	i	21%	1	A
		ii	15%	1	C
2	a		As the nitrate fertiliser increases from 0–160 kg ha⁻¹ the seed yield increases from 2.4 tonnes ha⁻¹ to 3.3 tonnes ha⁻¹ = **1**	2	CA
			And the percentage of oil decreases from 46% to 41% = **1**		
	b		37.5%	1	A
	c		0.025	1	A
	d		44%	1	A
	e		1:40	1	A
	f		40%	1	C
	g		Replicate field trials	1	C
3	a	i	Beginning of August until end of August increases from 4.0 to 4.5 g = **1**	2	CA
			End August until end November decreases from 4.5 to 1.5 g = **1**		
		ii	60%	1	A
		iii	More food/nectar available	1	C
	b	i	0.45 cm³ per gram per hour	1	A
		ii	7.5 cm³ per gram per hour	1	A
		iii	Birds do not enter torpor if the air temperature is above 20 °C	1	C
	c		21 cm³	1	A

Experimental questions

Question			Expected answer	Mark	Demand
1	a		Temperature **OR** pH of gel **OR** Type/concentration of stain **OR** Voltage applied to gel **(Any 2)**	2	CA
	b		Repeat the procedure with each sample	1	C
	c		32 mm	1	C
	d		Axes labelled (using exact headings provided in the table) and scale; points plotted accurately and joined with straight lines	1	C
	e		The smaller/fewer base pairs in the fragment, the further it travels in the gel **OR** converse	1	C
	f		11	1	C
	g		Donor 2	1	C
2	a		0.9 mm/min	1	A
	b	i	Oxygen taken up by organism and CO_2 given out = **1** CO_2 absorbed by solution/potassium hydroxide causes liquid to move towards tube = **1**	2	CA
		ii	Same mass of seeds and woodlice	1	C
		iii	Same apparatus without the living organism/with organism replaced with glass beads = **1** To show that oxygen uptake/movement of liquid was due to respiration/living organism = **1**	2	CA
		iv	Water bath	1	C
		v	To return the liquid drop to 0/start of scale (to run another experiment)	1	C
3	a	i	Light intensity	1	C
		ii	Temperature Concentration of *Chlorella* Concentration of CO_2 pH **(Any 1)**	1	C
	b		To allow time for *Chlorella* to acclimatise/to allow time for *Chlorella* to be carrying out photosynthesis under the experimental conditions/ To allow time for photosynthesis to take place/To allow time for intermediates to be produced **(Any 1)**	1	A

Question		Expected answer	Mark	Demand
	c	Repeat the experiment with a further 500 cm³ of *Chlorella* and obtain an average	1	C
	d	Correct scales **AND** labels [time (minutes) on *x*-axis and RuBP concentration (units) on *y*-axis] = **1** Correct plots **AND** straight line connections = **1**	2	CA
	e	In the light/when illuminated the RuBP concentration is low/8 units and the 3PG concentration is high/20 units = **1** When the light is switched off/in the dark the 3PG concentration decreases and the RuBP concentration starts to increase = **1**	2	CA

Mini extended response questions

Question	Expected answer	Mark	Demand
1	**1** Stem cells are unspecialised/undifferentiated cells (in animals) **2** That can divide/self-renew and/or differentiate **3** Differentiation is the process by which a cell expresses certain genes to produce proteins characteristic of that type of cell **4** There are tissue and embryonic stem cells **5** Embryonic stem cells differentiate into all cell types/are pluripotent **6** Tissue stem cells give rise to more limited cell types/are multipotent **7** Tissue stem cells are involved in growth and repair/renewal of the cells found in that tissue **(Any 5)**	5	CCCCA
2	**1** Genome is the sequence of bases in DNA **2** Sequences that code for proteins are called genes **3** Genes are coding sequences **4** Non-coding sequences do not code for proteins **5** Some non-coding sequences regulate transcription **6** Some non-coding sequences are transcribed to RNA but not translated **7** Some non-coding sequences are transcribed to form rRNA and tRNA **(Any 5)**	5	CCCCA
3	**1** Phylogenetics involves the study of gene sequences **2** Gene sequences are used to show evolutionary relatedness **3** Evolutionary relatedness is the basis of phylogenetic trees **4** To confirm timescales for phylogenetic trees, fossils are needed **(Any 3)** **5** Molecular clocks are graphs based on sequence differences of a particular protein or gene **6** Differences in sequences related to a protein or gene in different species are graphed on one axis **7** The timescale of divergence based on relative sequence differences is graphed on the other axis **8** Molecular clocks assume a constant rate of mutation **(Any 2/3)**	5	CCCCA

Question	Expected answer	Mark	Demand
4	**1** Competitive inhibitors have a similar shape to the substrate molecule **2** Competitive inhibitors bind to active sites of enzymes **3** They reduce reaction rates by blocking the active site **4** The effect is reduced by increasing the concentration of substrate **5** Non-competitive inhibitors bind to sites other than the active site **6** They alter the structure/shape of the active site **7** Their effect is not/little changed by increasing the substrate concentration **(Any 5)**	5	CCCCA
5	**1** Conformers' metabolism/metabolic rate/internal environment is dependent on their surroundings/external environment **2** Conformers use behaviour to maintain optimum metabolic rate **3** Regulators can maintain/regulate their metabolism/ metabolic rate/ internal environment regardless of external conditions **4** Regulators require energy for homeostasis/negative feedback **5** Conformers have narrower ecological niches **OR** regulators have a wider ecological niche **6** Conformers have lower metabolic costs/rates of metabolism **OR** regulators have higher metabolic costs/rates of metabolism **(Any 5)**	5	CCCCA
6	**1** Recombinant plasmids contain restriction sites that contain target sequences of DNA where specific restriction endonucleases cut **2** Regulatory sequences control gene expression **3** Origin of replication allows self-replication of the plasmid **4** Selectable markers such as antibiotic resistance genes protect the microorganism from a selective agent (antibiotic) that would normally kill it or prevent it growing **5** Selectable marker genes present in the vector ensure that only microorganisms that have taken up the vector grow in the presence of the selective agent (antibiotic) **6** As a safety mechanism, genes are often introduced that prevent the survival of the microorganism in an external environment **(Any 5)**	5	CCCCA
7	Food security: **1** Production of sufficient quantities of food **2** Production of sufficient quality of food **3** Accessibility/ability to distribute/spread food through the population **4** Knowledge required to use food properly Sustainability: **5** Ability to guarantee food security over longer periods **6** Food production must not degrade the natural resources on which agriculture depends **(Any 5)**	5	CCCCA

Question		Expected answer	Mark	Demand
8		**1** Plant field trials are carried out (in a range of environments) to compare the performance of different cultivars/treatments	5	CCCCA
		2 In designing field trials, account has to be taken of the selection of treatments, the number of replicates and the randomisation of treatments		
		3 The selection of treatments must ensure valid comparisons between cultivars		
		4 The number of replicates involved must take account of the variability within the sample		
		5 The randomisation of treatments is needed to eliminate bias when measuring treatment effects		
		(1 mark each)		
9		**1** Biodiversity is measured in terms of species/genetic/ecosystem diversity **(Any 2 = 1)**	5	CCCCA
		2 A third example from point 1		
		3 Species diversity is species richness and relative abundance/proportion of each species		
		4 Species richness is the number of different species		
		5 Genetic diversity is the number and frequency of different alleles in a population/species		
		6 Ecosystem diversity is the number of distinct ecosystems in an area/environment		
		(Any 5)		

Full extended response questions

Question			Expected answer	Mark	Demand
1	a		**1** Prokaryotes have circular chromosomes and plasmids	5	CCCCA
			2 Yeast has plasmids		
			3 Circular chromosomes in mitochondria/chloroplasts		
			4 Linear chromosomes in nucleus of eukaryotes		
			5 Prokaryotes have circular DNA **AND** eukaryotes have linear DNA (Only if point 1 or 4 not awarded)		
			6 Linear/eukaryotic/nuclear chromosome/DNA (tightly) coiled		
			7 Linear/eukaryotic/nuclear chromosome/DNA packaged with/wrapped around histones		
			(Any 5)		
	b		**1** The DNA is heated to 92–98 °C	4	CCCA
			2 The high temperature denatures the DNA/breaks hydrogen bonds which separates the strands		
			3 Temperature is cooled/lowered to 50–65 °C		
			4 Cooling allows primers to bind to the complementary target sequences		
			5 The temperature is raised to 70–80 °C		
			6 When heat-tolerant DNA polymerase is used to add DNA nucleotides to 3' end/primer		
			7 Repeated cycles allow millions of copies of the target sequence to be produced		
			(Any 4)		

Question		Expected answer	Mark	Demand
2	a	**1** Isolation barriers prevent gene flow between populations/populations interbreeding **2** Geographical isolation leads to allopatric speciation **3** Behavioural isolation leads to sympatric speciation **4** Ecological isolation leads to sympatric speciation **5** Different mutations occur on each side of isolation barrier **6** Different mutations may be selected on each side of the barrier **(Any 5)**	5	CCCCA
	b	**1** Natural selection is non-random increase in frequency of genetic sequences that increase survival **2** Any two from disruptional/directional/stabilising selection **3** Third type of selection from point 2 **4** After many generations/long period of time **5** If populations can no longer interbreed to produce fertile young then different species have formed/speciation has occured **(Any 4)**	4	CCCA
3		**1** Electron transport chain takes place on the inner membrane of the mitochondria/cristae **2** Electron transport chain is a collection of proteins attached to the membrane/cristae **3** NADH and FADH2 release the high-energy electrons to the electron transport chain (on the inner mitochondrial membrane/cristae) **4** Electrons flow down a chain of electron acceptors, releasing their energy **5** Energy is used to pump hydrogen ions (H^+) across the inner mitochondrial membrane **6** Return flow of the hydrogen ions (H^+) back into the matrix drives the enzyme ATP synthase **7** Synthesis of ATP from ADP + Pi **8** This stage produces most of the ATP generated by cellular respiration **9** Oxygen combines with hydrogen ions and electrons to form water **(Any 6)** **10** Regeneration of ATP from ADP and phosphate uses the energy released from cellular respiration **11** ATP is used to transfer the energy from cellular respiration to synthetic pathways/cellular processes (where energy is required) **12** Breakdown of ATP to ADP and phosphate/Pi releases energy **(Any 2)**	8	CCCCCCAA

Question	Expected answer	Mark	Demand
4	**1** Recombinant DNA technology involves the joining together of DNA molecules from two different species	10	CCCCCCCAAA
	2 Genes that remove inhibitory controls **OR** amplify specific metabolic steps can be introduced		
	3 This can increase the yield of a desired protein		
	4 Genes to prevent the survival of the microorganism in an external environment can be introduced as a safety mechanism		
	5 Marker genes can be introduced to improve selection		
	6 Recombinant plasmids/artificial chromosomes act as vectors to carry the DNA into host cell		
	7 Plasmid must contain origin of replication for self-replication		
	8 Restriction endonucleases cut target sequences from donor/source DNA (leaving sticky ends)		
	9 Treatment of donor chromosome and plasmid with the same restriction endonuclease		
	10 Complementary sticky ends are then combined using DNA ligase to form recombinant DNA		
	11 Modified microorganisms are cultured		
	12 Desired gene product/protein is collected/harvested		
	(Any 10)		
5	**1** Most crop plant pests are invertebrate animals such as insects/molluscs/nematode worms **(Any 2 = 1)**	9	CCCCCCCAA
	2 Plant diseases can be caused by fungi/viruses/bacteria **(Any 2 = 1)**		
	3 Pesticides include herbicides/insecticides/fungicides/molluscicides/nematicides **(Any 2 = 1)**		
	4 Selective herbicides have a greater effect on broad leaved plants		
	5 Systemic herbicides spread through the vascular system of plants (preventing regrowth) **OR** systemic insecticides/molluscicides/nematicides spread through the vascular systems of plants and kill the pests feeding on the plants.		
	6 Fungicide can be applied based on fungal disease forecasts		
	7 May be toxic to non-target species		
	8 May persist in the environment		
	9 May bioaccumulate in organisms		
	10 May biomagnify in food chains		
	11 May produce a selection pressure on a population **OR** could result in resistant populations		
	(Any 9)		

Question		Expected answer	Mark	Demand
6	a	**1** Social hierarchy is a rank order/pecking order in a group of animals **OR** Dominant/alpha **AND** subordinates/lower rank **2** Aggression/fighting/conflict/violence reduced **3** Ritualistic display/appeasement/threat/submissive behaviour **OR** Alliances formed to increase social status **4** Ensures best/successful genes/characteristics are passed on **OR** guarantees experienced leadership **(Any 3)** **5** Cooperative hunting is when animals hunt in a group/together **6** Increases hunting success **OR** allows larger prey to be brought down **OR** more successful than hunting individually **7** (Subordinate) animals all get more food/energy than by hunting alone **8** Less energy used/lost per individual **(Any 3)**	6	CCCCAA
	b	**1** Any two examples from bees, wasps, ants, termites **2** Only some members of colony (hive) reproduce/are fertile **OR** queen **AND** males/drones mate/reproduce **OR** only queen lays eggs **OR** some/most members of colony are sterile/infertile/do not reproduce **OR** some/most of colony are workers who are sterile **3** Any two examples of worker roles: raise relatives/defend hives/collect pollen/nectar/food/waggle dance to show direction of food **4** Social insects show kin selection/altruism between related individuals **5** Increases/helps survival of shared genes **OR** so shared genes are passed on to next generation **(Any 4)**	4	CCCA

Paper 1

Total marks: 25

Attempt ALL questions.

The answer to each question is either A, B, C or D. Decide what your answer is, then circle the appropriate letter.

There is only one correct answer to each question.

Allow yourself 40 minutes for Paper 1.

STUDENT MARGIN

1   The diagram shows part of a DNA molecule during replication. The bases are represented by numbers and letters.

Base 1 represents adenine and base 3 represents guanine.

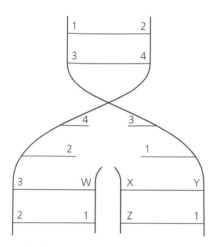

Which row in the table identifies bases W, X, Y and Z?

Applying KU

	W	X	Y	Z
A	cytosine	guanine	guanine	thymine
B	guanine	cytosine	cytosine	adenine
C	cytosine	guanine	cytosine	thymine
D	guanine	cytosine	guanine	adenine

2   The diagram represents part of the base sequence and direction of a strand of DNA. Which sequence and direction would be found on the complementary DNA strand?

Applying KU

5´ - - GAATTCGAT - - 3´

A   5´ - - GAATTCGAT - - 3´
B   5´ - - CTTAAGCTA - - 3´
C   3´ - - CUUAAGCUA - - 5´
D   3´ - - CTTAAGCTA - - 5´

3   Which of the following structures would be found in **both** prokaryotic and eukaryotic cells?

Demonstrating KU

A   ribosomes
B   mitochondria
C   chloroplasts
D   nuclei

4 Which of the following groups are **all** gene mutations?

A   insertion, deletion and substitution

B   substitution, duplication and translocation

C   translocation, insertion and deletion

D   deletion, duplication and translocation

5 The list shows different types of ribonucleic acid (RNA).

1   messenger RNA

2   transfer RNA

3   ribosomal RNA

Which type(s) of RNA is (are) transcribed from non-coding sequences of DNA?

A   1 only

B   1 and 2 only

C   2 and 3 only

D   1, 2 and 3

6 The genome of a cell contained 6×10^9 base pairs of which 4% coded for proteins. How many DNA codons coded for proteins?

A 8×10^7

B 8×10^8

C 2.4×10^7

D 2.4×10^8

7 The graph shows the results of an investigation in which the wing lengths of a population of house sparrows *Passer domesticus* in an area of North America measured in 1890 were compared with the wing lengths of a population in the same area in 1990.

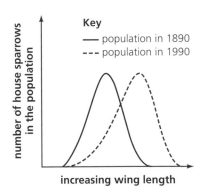

Which row in the table shows how selection pressure acts and the type of selection which is likely to have been involved in this example?

	Selection pressure	Type of selection
A	favours shorter wings	stabilising
B	acts against shorter wings	directional
C	favours longer wings	stabilising
D	acts against longer wings	directional

8 The diagram shows a possible phylogenetic tree for some groups of birds, most of which are flightless.

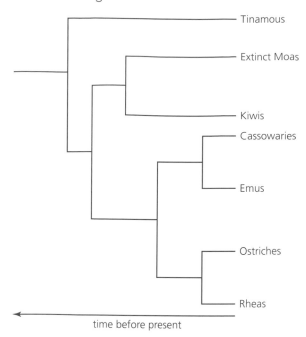

time before present

Which row in the table shows the species to which Kiwis are most and least closely related according to this phylogeny?

	Most closely related	Least closely related
A	Cassowaries	Rheas
B	Moas	Rheas
C	Cassowaries	Tinamous
D	Moas	Tinamous

9 The diagram shows an enzyme molecule (E) bound to its substrate (S) and a molecule (I) which can act as an inhibitor of the enzyme.

Which row in the table identifies the type of inhibitor shown and the effect on its action of an increase in substrate concentration?

	Type of inhibitor	Effect of an increase in substrate concentration
A	competitive	reduction of inhibition
B	non-competitive	reduction of inhibition
C	competitive	no effect on inhibition
D	non-competitive	no effect on inhibition

10 Equal volumes and concentrations of an enzyme and the substrate it breaks down were mixed with various concentrations of lead or calcium ions. The time taken for the complete breakdown of the substrate was measured and the results recorded in the table.

Metal ion concentration (units)	Time for complete substrate breakdown (seconds)	
	Lead ions	Calcium ions
0	42	42
2	47	21
20	380	49
200	1580	286

Which row in the table describes the effects of high concentrations of each metal ion on the activity of this enzyme?

	High concentration of lead ions	High concentration of calcium ions
A	promoted	promoted
B	promoted	inhibited
C	inhibited	inhibited
D	inhibited	promoted

Concluding

11 Three different mutated strains of yeast each lacked a different respiratory enzyme involved in the complete breakdown of glucose. The effect of the missing enzyme on each of the strains is shown in the table.

Strain	Effect of missing enzyme
P	cannot degrade pyruvate
Q	cannot synthesise pyruvate from intermediates
R	cannot convert citrate to oxaloacetate

Which of the strains could produce ethanol?

A   P only

B   R only

C   P and Q

D   Q and R

Applying KU

12 The list shows the roles of substances which make up the membranes of cells.

1   pump ions across the membrane using ATP

2   catalyse chemical reactions

3   act as pores to allow selected substances through

Which roles can be carried out by proteins?

A   1 and 2 only

B   1 and 3 only

C   2 and 3 only

D   1, 2 and 3

Demonstrating KU

13 Which row in the table shows the characteristics of the heart and circulatory system of a bird?

	Number of heart chambers		Degree of separation between oxygenated and deoxygenated blood
	Atria	Ventricles	
A	1	1	incomplete
B	1	2	complete
C	2	2	incomplete
D	2	2	complete

Demonstrating KU

14 The graph shows the results of an experiment to show the effect of air temperature on the metabolic rate of a lizard and a small mammal.

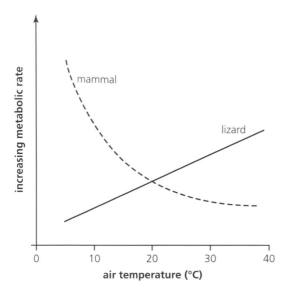

Which row in the table identifies the temperatures at which oxygen consumption will be the greatest in the tissues of each animal?

	Temperature (°C)	
	Lizard	Mammal
A	20	20
B	40	40
C	40	5
D	5	40

Applying KU

15 The list shows procedures which can be carried out when culturing wild strains of microorganisms in laboratory fermenters.

1   supply the culture with raw material for biosynthesis
2   expose the culture to UV light
3   add mutagenic chemicals to the culture
4   sterilise the fermenter before adding the culture

Which of these procedures could increase the chances of producing a new improved strain of microorganisms?

A   1 and 2 only
B   2 and 3 only
C   3 and 4 only
D   1 and 3 only

Demonstrating KU

16 Recombinant DNA technology involves modifying organisms by using vectors to carry foreign genetic information into their genomes.

In which circumstances are artificial chromosomes preferable to plasmids as vectors?

A   tiny fragments of DNA are to be inserted

B   regulatory sequences are to be inserted

C   large fragments of genetic information are to be inserted

D   selectable markers are to be inserted

Demonstrating KU

17 A chlorophyll extract was made from some plant leaves. Light was shone through the extract then through a spectroscope.

Which of the spectra would be seen through the spectroscope?

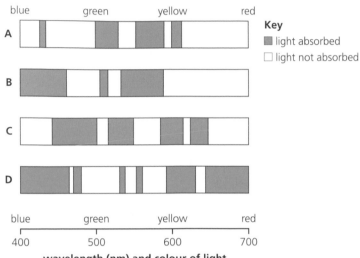

Predicting

18 In photosynthesis, the function of carotenoid pigments is to

A   receive light energy from chlorophyll for use in photolysis

B   allow plants to absorb a wider range of light wavelengths

C   increase photosynthesis at low light intensities

D   change the absorption spectrum of chlorophyll.

Demonstrating KU

19 Descriptions of behaviour sometimes observed in captive or domesticated animals are shown in the list.

1   preferential feeding in chimpanzees

2   feather grooming by hens

3   repeated foot stamping by bulls

4   stereotypic head flicking in pigs

Which behaviours could be considered indicators of poor welfare of the animals?

A   1 and 3 only

B   1 and 4 only

C   2 and 3 only

D   3 and 4 only

Applying KU

20 The Mandarin duck *Aix galericulata* was brought to Britain from East Asia in the eighteenth century as an ornamental species in wildfowl collections. Individuals have escaped into wild habitats around the country and in some places have become established in wild communities.

Which terms can be applied to species such as the Mandarin duck?

A   indigenous and invasive

B   introduced and naturalised

C   indigenous and naturalised

D   introduced and invasive

Applying KU

21 Livestock breeders carry out crossbreeding programmes.

Crossbreeding

   A   combines characteristics of separate breeds

   B   causes inbreeding depression in offspring

   C   maintains characteristics of a new breed

   D   increases the homozygosity of a breed.

22 Poppies are annual weeds found in cereal fields.

Which row in the table identifies the competitive adaptations expected in poppies?

	High seed output	Storage organs	Short life cycle
A	✓	✗	✓
B	✓	✓	✗
C	✗	✗	✗
D	✗	✓	✓

23 The main advantage gained by using land for crops rather than livestock is that

   A   crops produce more food per unit area of land than livestock

   B   crops produce food with higher nutritional values than livestock

   C   crop production requires fewer chemical applications than livestock production

   D   crop production creates less environmental damage than livestock.

24 Capsicum is a crop which contributes to food security in many parts of the world.

The graph shows the changes in the vitamin C content and the mass of the pigment chlorophyll in ripening capsicum fruits.

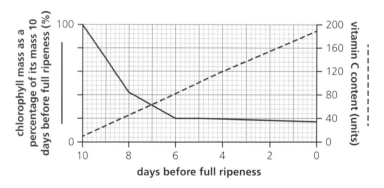

Which of the following is a valid conclusion which can be drawn from the data in the graph?

During the ripening of capsicum fruits

   A   chlorophyll is converted into vitamin C between 10 and 8 days before full ripeness

   B   50% of chlorophyll is broken down by 9 days before full ripeness

   C   vitamin C content increases steadily between 4 and 6 days before full ripeness

   D   vitamin C content equals chlorophyll content 7 days before full ripeness.

25 The graph shows the levels of two different pesticides which were applied to the commercial tomato plant crop in a large greenhouse between 1994 and 2012.

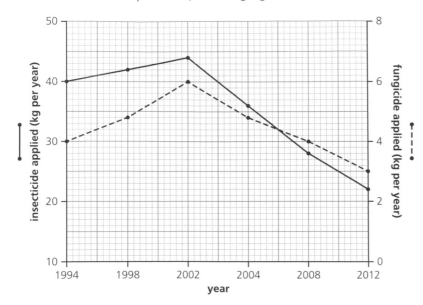

In which year was there the smallest difference between the levels of insecticide and fungicide applied?

A   2004

B   2006

C   2008

D   2012

Selecting

[End of Paper 1]

Paper 2

Total marks: 95

Attempt ALL questions.

Write your answers clearly in the spaces provided. If you need additional space for answers or rough work, please use separate pieces of paper.

Allow yourself 2 hours and 20 minutes for Paper 2.

	MARKS	STUDENT MARGIN

1   DNA and RNA molecules are found in the cells of both prokaryotes and eukaryotes.

    a   Give **two** structural differences between DNA and RNA molecules.   **2**   Demonstrating KU

        1 _____

        2 _____

    b   Describe the function of RNA polymerase in the synthesis of a primary RNA transcript.   **2**   Demonstrating KU

    c   In eukaryotic cells, mRNA is spliced after transcription.
        Describe what happens during RNA splicing.   **2**   Demonstrating KU

    d   Describe the role of mRNA in cells.   **1**   Demonstrating KU

2   The diagram shows details of a molecule of transfer RNA (tRNA).
    An area of the diagram has been enlarged.

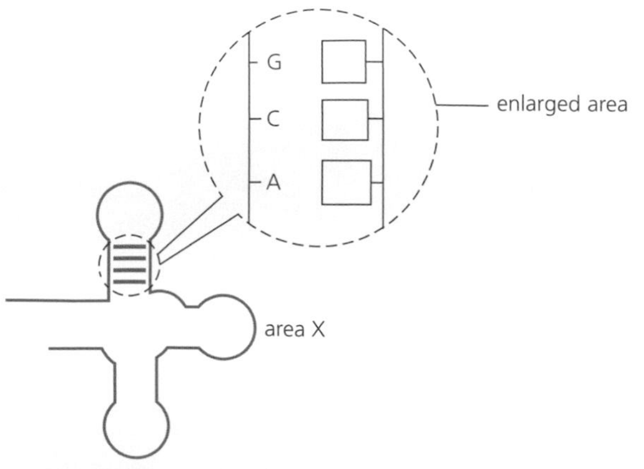

    a **i** Complete the boxes in the enlarged area of the diagram above by adding letters to represent the complementary RNA bases.   **1**   Demonstrating KU

        ii Use the letter **W** to label the position of the amino acid binding site **on the diagram**.   **1**   Applying KU

	MARKS	STUDENT MARGIN

b Area X on the tRNA molecule contains an anticodon.

Describe an anticodon and explain its importance in translation.

Description:

_____

_____

Importance: _____

_____

_____

2   Demonstrating KU

c tRNA is produced by transcription of a region of DNA carrying non-coding sequences.

Give **one** other function of non-coding sequences.

_____

_____

1   Demonstrating KU

3 The diagram shows three species of flycatcher and a representation of their breeding ranges in Europe. Their ranges overlap to some extent and interbreeding between Pied and Collared Flycatchers occasionally occurs in area H.

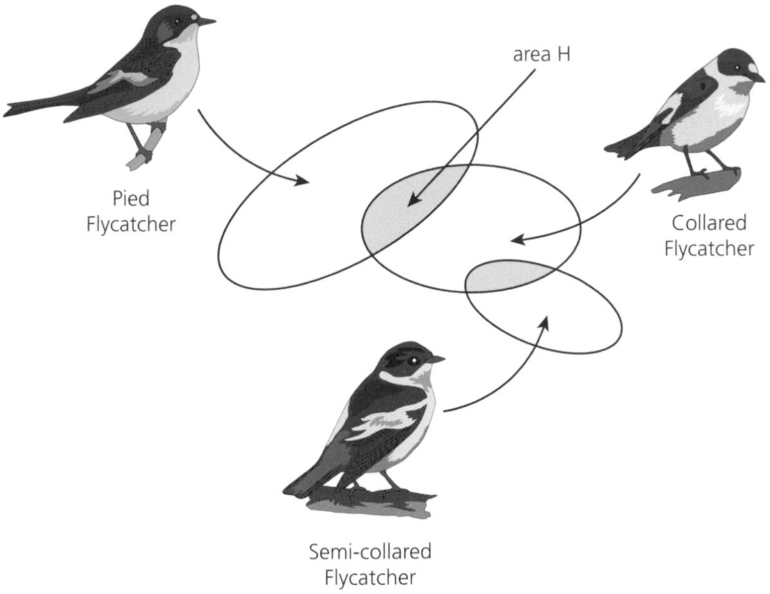

Pied Flycatcher

area H

Collared Flycatcher

Semi-collared Flycatcher

a The three species are thought to have evolved by allopatric speciation from a common ancestor.

Describe how allopatric speciation occurs.

_____

_____

_____

_____

3   Demonstrating KU

b Pied and Collared Flycatchers are considered to be different species even though they may occasionally interbreed in area H.

Describe the evidence which would confirm that, despite the ability to interbreed, Pied and Collared Flycatchers are separate species.

_____

_____

_____

1   Applying KU

4 The diagram shows a pair of homologous partner chromosomes exchanging genetic information during a process leading to the formation of four gametes in the testes of a male mammal.

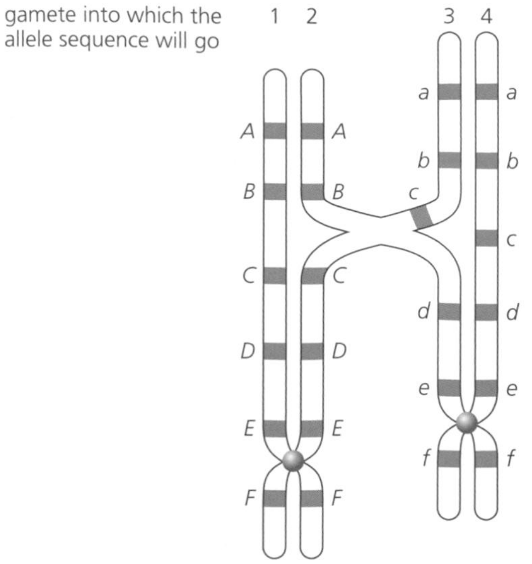

gamete into which the allele sequence will go

a The process shown will result in four gametes (1–4) being formed.

Complete the table using information from the diagram to show the sequence of alleles expected in gamete 2.

Gamete	Sequence of alleles in gametes
1	A B C D E F
2	
3	a b c C D E F
4	a b c d e f

1   Selecting

b Name the chromsome mutation, the results of which are visible in gamete 3, and describe the importance of that mutation in evolution.

3   Demonstrating KU

Name: _____

Importance: _____

_____

_____

_____

MARKS | STUDENT MARGIN

5 The polymerase chain reaction (PCR) is a technique used to amplify DNA samples in the laboratory. The diagram shows the contents of a tube being set up for PCR.

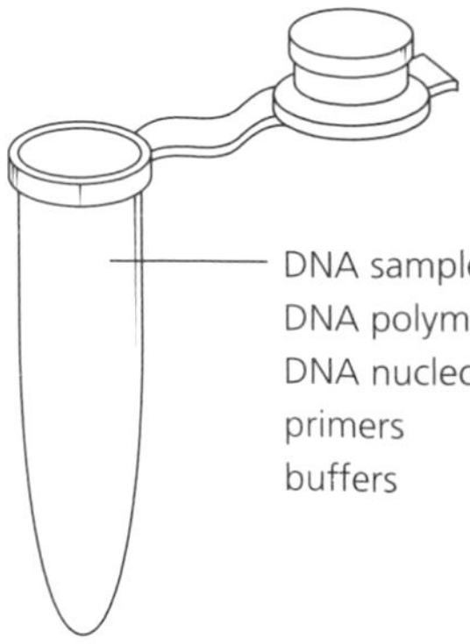

DNA sample to be amplified
DNA polymerase
DNA nucleotides
primers
buffers

a Describe the contents of a control tube designed to prove that the procedure being used is free of contaminating DNA.

1   Planning

_____

_____

b During the first stage of the PCR process, the tube will be heated to 95°C. Explain why the DNA polymerase remains active even after exposure to this temperature.

1   Applying KU

_____

_____

c Describe the role of buffers in biological experiments.

1   Planning

_____

_____

d Describe a practical situation in which DNA needs to be amplified.

1   Demonstrating KU

_____

_____

6 During respiration in yeast, hydrogen ions are released from glucose molecules.

In experiments, hydrogen ions can decolourise the indicator solution methylene blue. The faster the rate of respiration, the faster methylene blue will decolourise.

The effect of temperature on the rate of respiration in yeast was measured using the following method.
- A water bath at 15°C was set up.
- Four tubes containing the substances shown in the table were placed in the water bath.

Tube	Contents
1	2 cm³ yeast suspension
2	2 cm³ glucose solution
3	2 cm³ methylene blue solution
4	empty

- The tubes were left in the water bath for 10 minutes, then their contents were mixed into tube 4 and the time taken for the methylene blue to decolourise was measured.

• The procedure was repeated at a range of temperatures and the results are shown in the table.

Temperature of water bath (°C)	Time for methylene blue to decolourise (s)
15	140
25	90
30	60
35	30
45	80

a On the grid, plot a line graph of the results.

2   Presenting

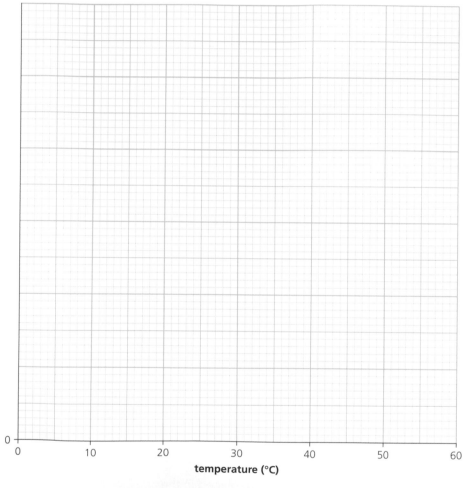

temperature (°C)

b i Identify the dependent variable in this experiment.

1   Planning

ii Identify **one** variable which must be kept constant at each temperature to allow valid conclusions to be drawn.

1   Planning

c i Describe the contents of a control tube which should be set up at each temperature to ensure that the decolourisation reaction was due to respiration.

1   Evaluating

		MARKS	STUDENT MARGIN

ii Explain why the four tubes were kept in the water bath for 10 minutes **before** their contents were mixed.

_____

_____

1 — Planning

d Explain the pattern of results obtained when the temperature of the water bath was raised from 35 °C to 45 °C.

_____

_____

1 — Applying KU

7 Some species of pine tree are adapted to tolerate the adverse conditions of extremely low winter temperatures. High levels of sucrose accumulate in the sap of their needles as winter progresses to prevent them freezing solid.

The graph shows how the average concentration of sucrose in samples of needles of two species of pine tree *Pinus strobus* and *Pinus virginiana* changed over a two-year period. The sucrose concentration in the needles was measured at the beginning and the end of April, August and December in each year.

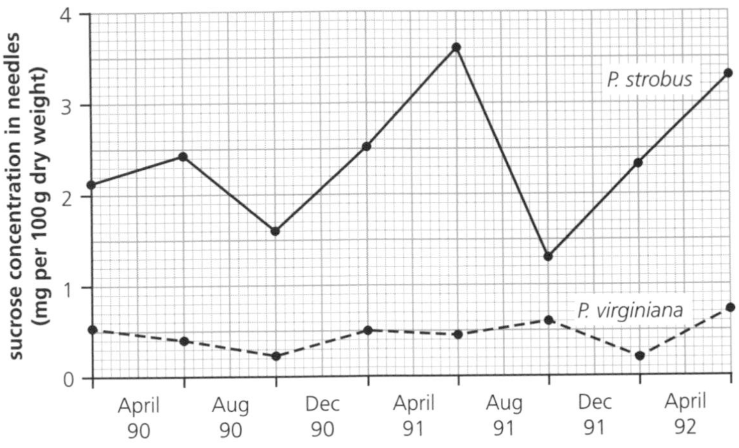

a **i** Calculate the percentage increase in the average sucrose content in the *P. strobus* needles between the end of August 1990 and the end of April 1991.

_____ %

1 — Processing

ii Calculate the difference between the sucrose content of *P. strobus* and *P. virginiana* needles at the end of December 1990.

_____ mg per 100g dry weight

1 — Processing

b Identify evidence in the data which suggests that *P. strobus* can adjust its accumulation of sucrose depending on the winter temperatures.

_____

_____

2 — Concluding

		MARKS	STUDENT MARGIN

c Predict how winter temperatures in the range of *P. strobus* compare with those in the range of *P. virginiana* and give evidence from the graph which would support your prediction.

2 — Predicting

_____

_____

d Give an example of how an animal species can **avoid** adverse conditions which occur regularly.

1 — Demonstrating KU

_____

_____

8 The diagrams represent the hearts of the three different vertebrate groups shown.

fish   amphibian   mammal

a **i** Fish have a single circulation.
Explain what is meant by a single circulation.

1 — Demonstrating KU

_____

_____

ii Amphibians have a double circulation.
Explain the advantage to amphibians of having a double circulation.

1 — Demonstrating KU

_____

_____

b Add arrows to the diagram in blood vessels X and Y in the mammal heart to show the direction of blood flow within them.

1 — Applying KU

c Explain how structure Z allows mammals to maintain high metabolic rates.

2 — Applying KU

_____

_____

	MARKS	STUDENT MARGIN

9 Mitochondria have double membranes separated by a space, as shown in the diagram. Electron transport chains are located along the inner membranes.

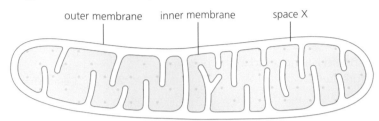

outer membrane   inner membrane   space X

a Name the substance that the carriers in the transport chains are made from.

1 — Demonstrating KU

b Explain how a high concentration of hydrogen ions can be built up in space X shown in the diagram.

1 — Demonstrating KU

c Describe how ATP is generated in the electron transport chains.

2 — Demonstrating KU

10 The diagram shows a section through a fermenter set up to grow a culture of genetically modified bacteria.

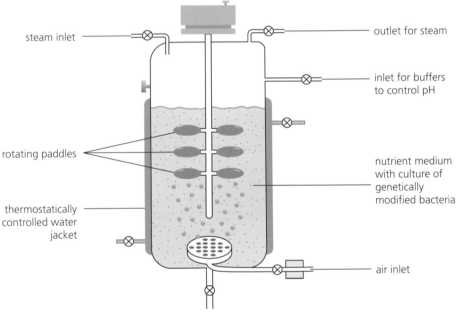

steam inlet

outlet for steam

inlet for buffers to control pH

rotating paddles

nutrient medium with culture of genetically modified bacteria

thermostatically controlled water jacket

air inlet

a i State why fermenters such as this have steam pumped through them before the culture of microorganisms is added.

1 — Demonstrating KU

ii Suggest why the rotating paddles are needed in the fermenter.

1 — Applying KU

	MARKS	STUDENT MARGIN

b Explain why controlling the temperature and pH of the fermenter is important.

_____

_____

MARKS: 1 — Demonstrating KU

c Give **one** example of an ingredient which should be included in the nutrient medium **and** explain the function of this ingredient.

Ingredient: _____

Function: _____

_____

MARKS: 2 — Demonstrating KU

11 Papaya Ringspot Virus (PRSV) threatens food security because it damages papaya, a tropical fruit. Papaya plants with resistance to this virus have been produced using recombinant DNA technology as shown in the flow diagram.

Step 1
Plasmids removed from the bacterium *Agrobacterium tumefaciens* and modified with PRSV resistance genes and selectable markers.

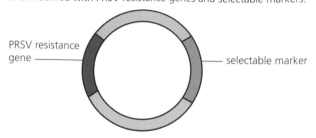

PRSV resistance gene

selectable marker

Step 2
Modified plasmids returned to *Agrobacterium tumefaciens* cells and a selective agent used to identify successfully modified bacteria.

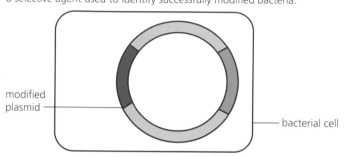

modified plasmid

bacterial cell

Step 3
Successfully modified bacteria cultured with papaya cells and PRSV resistance genes taken into the genome of papaya cells.

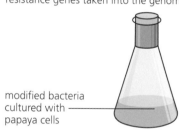

modified bacteria cultured with papaya cells

Step 4
Papaya cells grown on agar plate to develop PRSV-resistant papaya plantlets.

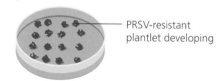

PRSV-resistant plantlet developing

a i Describe how isolated resistance genes can be added into unmodified plasmids in **Step 1**.

_____

_____

MARKS: 1 — Applying KU

ii Describe the role of the selectable marker in **Step 2**.

_____

_____

MARKS: 1 — Applying KU

b Explain why yeast cells are often used in preference to bacterial cells to receive recombinant DNA.

_____

_____

MARKS: 1 — Demonstrating KU

c Bacteria release and take up plasmids naturally.
Give the term used to describe this type of gene transfer.

_____

MARKS: 1 — Demonstrating KU

12 Weed plants compete with crop plants and reduce their productivity. There
    have been changes in herbicide usage in attempts to control weeds and so to
    increase crop productivity. There is concern that changes in herbicide use have
    had unintended impacts on the environment.

    Graph 1 shows data relating to applications of herbicide to farmland in a part of
    the UK between 1970 and 2000.

    Graph 2 show the change in the total populations of 19 species of farmland
    birds over the same period of time as percentages of the 1970 populations.

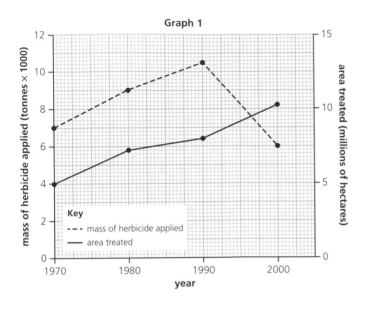

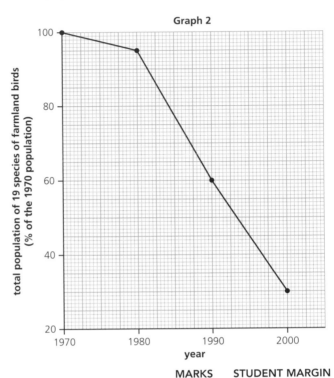

	MARKS	STUDENT MARGIN
a i **Use values from Graph 1** to describe the changes in mass of herbicide applied between 1970 and 2000.	2	Selecting
ii Calculate the percentage increase in mass of herbicide applied between 1970 and 1990.	1	Processing
iii Calculate the average increase per year in area treated between 1970 and 2000.	1	Processing
iv Calculate the mass of herbicide applied per million hectares in 1990.	1	Processing

a i **Use values from Graph 1** to describe the changes in mass of herbicide
      applied between 1970 and 2000.

      _____

      _____

   ii Calculate the percentage increase in mass of herbicide applied between
       1970 and 1990.

       _____ %

   iii Calculate the average increase per year in area treated between 1970
        and 2000.

        _____ hectares

   iv Calculate the mass of herbicide applied per million hectares in 1990.

        _____ tonnes

			MARKS	STUDENT MARGIN

b **i** From **Graph 2**, identify the year by which the total population of farmland bird species had declined by 40%.

_____

1 — Selecting

ii Give the percentage of the 1970 population which would have been predicted to remain by 2002 based on this data.

_____ %

1 — Predicting

c From the data given, suggest a hypothesis to explain the decline of farmland bird populations.

_____

_____

1 — Concluding

d What evidence is there in the data given that steps were taken during the period to lessen the impact of herbicides on the environment?

_____

_____

1 — Concluding

13 Answer **either A or B** in the space below.
Labelled diagrams may be used where appropriate.
A Describe the measurable components of biodiversity.
OR
B Describe the effects of invasive species on biodiversity.

_____

_____

_____

_____

4 — Demonstrating KU

14 **a** Integrated pest management (IPM) uses a combination of types of pest control.
Complete the table to show the different types of pest control and examples of their use.

3 — Demonstrating KU

Type of pest control	Example of use
	applying insecticide to crops
biological	
cultural	

b State **one** environmental problem associated with applying insecticide to crops.

_____

_____

_____

1 — Demonstrating KU

MARKS	STUDENT MARGIN

15 a In an investigation into a sequence of metabolic reactions, a culture of the green alga *Chlorella* was kept in the dark for 24 hours. The culture was then exposed to light and samples of the algal cells obtained at four time intervals over 30 seconds of illumination.

Substances present in extracts made from each sample were separated using chromatography techniques and the results are shown in the diagram.

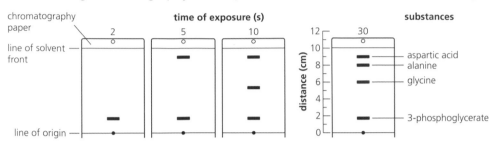

i **Using information from the diagram**, complete the metabolic pathway to show the sequence in which the substances were produced.

1 Concluding

_____ → _____ → _____ → _____

ii The relative front value (Rf) of a substance can be calculated using the following formula:

1 Processing

$$Rf = \frac{\text{distance travelled by substance from origin}}{\text{distance travelled by solvent from origin}}$$

Use the formula to calculate the Rf value for alanine.

Rf value for alanine = _____

b RuBP is a substance which acts as the carbon dioxide acceptor in carbon fixation. The graph shows the relative concentrations of RuBP and 3-phosphoglycerate in the algal cells when they were exposed to light and dark periods.

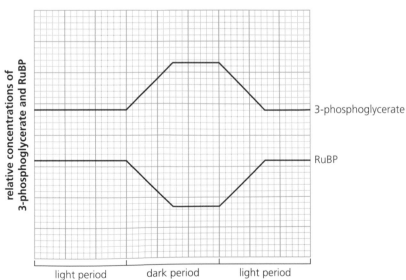

i Give evidence from the graph which supports the hypothesis that 3-phosphoglycerate can be converted to RuBP in light conditions.

1 Applying KU

_____

_____

		MARKS	STUDENT MARGIN

ii During photosynthesis, 3-phosphoglycerate combines with hydrogen and is phosphorylated by ATP to produce an intermediate substance which is then converted to RuBP.

Name the intermediate substance involved **and** name the photosynthetic reaction which provides the hydrogen.

2   Demonstrating KU

Substance: _____

Reaction: _____

16 The clownfish *Amphiprion ocellaris* lives among the stinging tentacles of the sea anemone *Heteractis magnifica*. The clownfish is highly territorial and drives off other fish that would eat the anemone.

The stinging tentacles of the anemone, which are harmless to the clownfish, help protect the clownfish from predators.

a Explain why this example can be considered as mutualism.

1   Applying KU

_____

_____

b i Describe how each species in a parasitic relationship is affected.

1   Demonstrating KU

_____

_____

ii Explain why a parasite is unlikely to survive out of contact with its host.

1   Demonstrating KU

_____

_____

iii Give **one** method by which a parasite can be transmitted to a new host.

1   Demonstrating KU

_____

17 Chimpanzees are primates. Most of their long period of parental care is carried out by their mothers.

The table shows the results of a study into the average age of offspring and descriptions of maternal contact behaviour in a large group of chimpanzees.

Average age of offspring (years)	Maternal contact behaviour
0–0.5	carried in chest-to-chest contact with mother
0.6–2.0	carried on mother's back
2.1–3.2	travelling and sitting up to 5 metres from mother
3.3–8.0	venturing more than 5 metres from mother
greater than 8.0	remaining close but playing independently and interacting socially with other group members

a i Calculate the percentage of its life that an average 2.8-year-old chimp has spent being carried on its mother's back.

1   Processing

_____ %

ii Describe the difference in maternal contact behaviour which would be expected between a young chimpanzee of 2.9 years and another of 3.6 years.

1   Selecting

_____

_____

	MARKS	STUDENT MARGIN

b i Describe why the long period of parental care is an advantage to chimpanzees.

_____

_____

1 — Applying KU

ii Give **two** examples of behaviours which have evolved to support social hierarchy in primates.

_____

_____

2 — Demonstrating KU

18 Answer **either question A or B** in the space below.

Labelled diagrams may be used where appropriate.

A Give an account of stem cells under the following headings:

 i types of stem cell and their properties — **5** — Demonstrating KU

 ii the research and therapeutic uses of stem cells. — **3** — Demonstrating KU

OR

B Give an account of the use of genomic sequencing in:

 i the determination of the sequence of events in evolution — **3** — Demonstrating KU

 ii personal genomics and health. — **5** — Demonstrating KU

_____

_____

_____

_____

_____

_____

_____

_____

_____

_____

_____

_____

_____

_____

_____

_____

_____

_____

_____

_____

_____

_____

_____

[End of Practice Exam A]

Paper 1

Total marks: 25

Attempt ALL questions.

The answer to each question is either A, B, C or D. Decide what your answer is, then circle the appropriate letter.

There is only one correct answer to each question.

Allow yourself 40 minutes for Paper 1.

STUDENT MARGIN

1 Which row in the table shows the organisation of DNA in the nuclei and mitochondria of a eukaryotic animal cell?

Demonstrating KU

	Nucleus	Mitochondrion
A	circular chromosomes	circular chromosomes
B	linear chromosomes	plasmids
C	linear chromosomes	circular chromosomes
D	circular chromosomes	plasmids

2 A fragment of DNA was found to consist of 80 base pairs.

What is the total number of deoxyribose sugars in this fragment?

Applying KU

A 20

B 40

C 80

D 160

3 Which of the following statements related to cell differentiation is correct?

Demonstrating KU

A Meristems are regions of differentiated cell types in plants.

B Differentiated cells express genes to produce proteins characteristic of that cell type.

C Embryonic stem cells can differentiate into a limited range of cell types.

D Adult tissue stem cells can differentiate into all cell types.

4 Coding sequences of DNA in the human genome are transcribed to produce

Demonstrating KU

A mRNA only

B tRNA only

C mRNA and tRNA

D mRNA, tRNA and rRNA.

5 The graph shows the changes in number of human stem cells in a culture. The activity of the enzyme glutaminase present in the cells over an 8-day period is also shown.

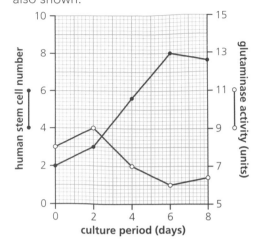

What was the human stem cell number when the glutaminase activity was at its maximum?

A 2   B 3   C 4   D 9

6 Types of single gene mutation are given in the list.

1 substitution

2 insertion

3 deletion

Which of these would have frameshift effects on the polypeptide produced?

A 1 only

B 1 and 2 only

C 1 and 3 only

D 2 and 3 only

7 One example of a molecular clock compares the amino acid sequences of the protein cytochrome C from different animal species. The differences in amino acid sequence were caused by mutations of the gene encoding cyctochrome C which have occurred in the past.

The clock is based on an assumption that mutations in this gene occur

A randomly

B non-randomly

C at a varying rate

D at a constant rate.

8 This reaction is part of a metabolic pathway in cells:

polypeptides → amino acids

Which row in the table identifies the type of reaction and whether it releases or takes up energy?

	Type of reaction	Energy released or taken up
A	catabolic	released
B	anabolic	released
C	catabolic	taken up
D	anabolic	taken up

9 Peppered moths *Biston betularia* are predated by birds which locate them resting on surfaces during day time. The moth has both light-winged and dark-winged forms.

In a study into selection pressures, equal numbers of light and dark-winged forms of peppered moth, each marked with a spot of paint, were released into two different areas. After several days, efforts were made to recapture the moths which had not been predated and the results are shown in the chart.

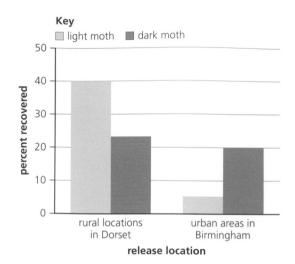

Which of the following is supported by the data in the chart?

A   Light moths had a selective advantage in urban Birmingham.

B   Dark moths had a selective disadvantage in urban Birmingham.

C   Dark moths had a selective advantage in rural Dorset.

D   Light moths had a selective advantage in rural Dorset.

10 Using recombinant DNA technology, the bacterium *Escherichia coli* can be modified so that it can produce human growth hormone (HGH).

The following steps are involved.

1   Culture large quantities of *E. coli* in nutrient medium.

2   Insert HGH gene into *E. coli* plasmid DNA.

3   Cut HGH gene from human chromosome using enzymes.

4   Extract HGH from culture medium.

The correct order of these steps is

A   1, 4, 3, 2

B   1, 2, 3, 4

C   3, 2, 1, 4

D   3, 1, 2, 4

Selecting

Demonstrating KU

11 An investigation was carried out into the uptake of sodium ions by animal cells. The graph shows the rates of sodium ion uptake and breakdown of glucose at different concentrations of oxygen.

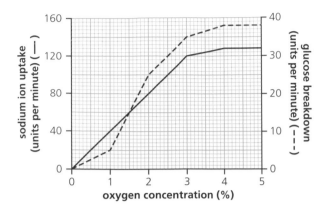

What was the number of units of glucose broken down **over a 5-minute period** when the concentration of oxygen in solution was 2%?

A   20

B   25

C   100

D   125

Processing

12 The diagram shows a metabolic pathway that is controlled by end product inhibition.

metabolite 1

enzyme P ↓

metabolite 2

enzyme Q ↓

metabolite 3

enzyme R ↓

metabolite 4

For metabolite 4 to bring about end product inhibition of the overall pathway, with which of the following will it interact?

A   metabolite 1

B   metabolite 3

C   enzyme P

D   enzyme R

Applying KU

13 The West African lungfish lives in a habitat in which hot, dry conditions can occur at any time of the year. To survive these conditions, the lungfish respond by burrowing into the mud, enclosing themselves in a cocoon of mucus and becoming dormant.

Which of the following descriptions applies to this type of dormancy?

A   predictive aestivation

B   consequential aestivation

C   predictive hibernation

D   consequential hibernation

Applying KU

14 The diagram shows a bacterial cell that has been magnified 700 times.

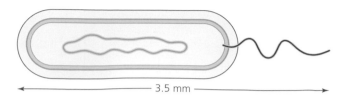

3.5 mm

What is the actual size of this cell in micrometers (μm)?

A   0.005

B   0.05

C   0.5

D   5.0

Processing

15 The diagram illustrates the circulatory system of a fish. The arrows indicate the direction of blood flow.

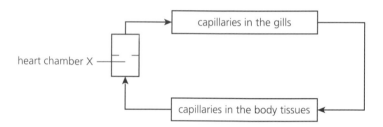

capillaries in the gills

heart chamber X

capillaries in the body tissues

Which row in the table identifies chamber X and the relative blood pressures in capillaries in gills and body tissues?

Applying KU

	Chamber X	Blood pressure in the capillaries in the gills	Blood pressure in the capillaries in the body tissues
A	atrium	higher	lower
B	ventricle	lower	higher
C	atrium	lower	higher
D	ventricle	higher	lower

16 The action spectrum of photosynthesis is a measure of the ability of plants to

A   absorb different wavelengths of light

B   absorb light of different intensities

C   extend the range of wavelengths of light absorbed

D   use light of different wavelengths in photosynthesis.

Demonstrating KU

17 The following statements refer to photosynthesis.

1   Carbon dioxide is fixed by the enzyme RuBisCO.

2   Water is split into hydrogen and oxygen.

3   G3P is used to regenerate RuBP.

Which of the statements refer to the carbon fixation stage?

A   1 and 2 only

B   1 and 3 only

C   2 and 3 only

D   1, 2 and 3

Applying KU

18 In self-pollinating plant species such as barley, natural selection reduces inbreeding depression by eliminating

    A   deleterious alleles

    B   recessive alleles

    C   heterozygous alleles

    D   dominant alleles.

Demonstrating KU

19 European bison were hunted to near extinction in the early twentieth century. The animals living today are all descended from 12 surviving individuals. Their population has extremely low genetic variation, which may be beginning to affect the reproductive ability of bulls.

    Which term is used to refer to the loss of genetic variation associated with such a serious decline in population?

    A   directional selection

    B   bottleneck effect

    C   overexploitation

    D   inbreeding depression

Applying KU

20 Which of the following is an example of a cultural method of protecting crops from pests?

    A   spraying with pesticide

    B   introducing a natural predator of the pest

    C   applying fungicide

    D   using crop rotation

Demonstrating KU

21 Which row in the table best describes the effects of altruistic behaviour on the donor and the recipient?

	Effect on donor	Effect on recipient
A	harmed	harmed
B	benefits	harmed
C	harmed	benefits
D	benefits	benefits

Demonstrating KU

22 A species of South American ant inhabits the thorns of a species of *Acacia* tree. The ants receive nectar and shelter from the plant. The ants protect the plant from attack by other insects.

    This is an example of

    A   parasitism

    B   mutualism

    C   kin selection

    D   predation.

Applying KU

23 Which row in the table shows the roles of the different members of a honey bee hive?

	Members of hive		
	Queen	Drone	Worker
A	reproductive female	non-reproductive female	reproductive male
B	non-reproductive female	reproductive male	reproductive female
C	reproductive female	reproductive male	non-reproductive female
D	non-reproductive female	reproductive female	reproductive male

Demonstrating KU

24 The table shows the number of plants of different species found in a single quadrat sample taken during a study of species diversity on a Scottish moorland site.

Species	Number of plants in sample (n)	$n \times (n - 1)$
Calluna vulgaris	6	30
Erica tetralix	3	6
Erica cinerea	2	2
Nardus stricta	1	0
Molinia caerulea	4	12
Total number of species in sample (N) = 5		Total $n \times (n - 1)$ = 50

Simpson's index of biodiviersity (D) is calculated by the following equation.

$$D = \frac{\text{Total } n \times (n - 1)}{N \times (N - 1)}$$

What is D for this quadrat?

A 0.4

B 2.5

C 10

D 250

25 The diagram shows the results of destruction of habitat to leave an undestroyed core area and four fragments around its edge.

Which remaining fragment is likely to exhibit the lowest species biodiversity?

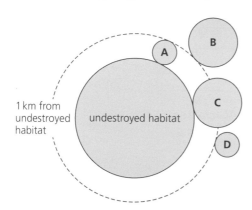

1 km from undestroyed habitat

undestroyed habitat

[End of Paper 1]

Processing

Applying KU

Paper 2

Total marks: 95

Attempt ALL questions.

Write your answers clearly in the spaces provided. If you need additional space for answers or rough work, please use separate pieces of paper.

Allow yourself 2 hours and 20 minutes for Paper 2.

	MARKS	STUDENT MARGIN

1 a The diagram shows a stage in the replication of a DNA molecule.

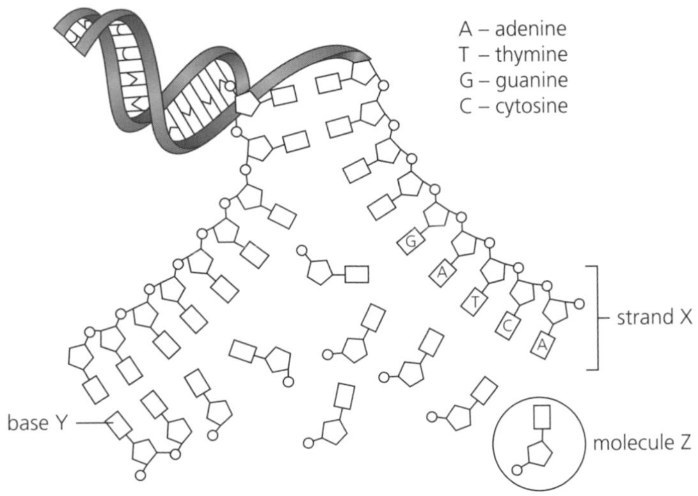

A – adenine
T – thymine
G – guanine
C – cytosine

strand X

base Y

molecule Z

 i   Identify the feature, shown in the diagram, which confirms that strand X is the lagging strand.

_____

1 Applying KU

 ii   Name base Y.

_____

1 Applying KU

 iii   Name molecule Z.

_____

1 Applying KU

b The polymerase chain reaction (PCR) is a laboratory technique which uses thermal cycles to amplify DNA.

 i   Give the temperature in a thermal cycle at which strands of DNA are separated from each other.

_____ °C

1 Demonstrating KU

 ii   Calculate how many molecules of DNA would be produced after a single molecule of DNA went through five complete thermal cycles of PCR.

_____ molecules

1 Applying KU

	MARKS	STUDENT MARGIN

iii Name the technique used to separate fragments of DNA made from an amplified sample to produce a unique DNA profile for an individual.

1 — Demonstrating KU

iv Give an example of a practical use of being able to identify an individual from their DNA profile.

1 — Demonstrating KU

2 Many genes have their coding DNA split by sections of DNA which do not code for protein.

The diagram shows the steps involved in the expression of the gene for insulin.

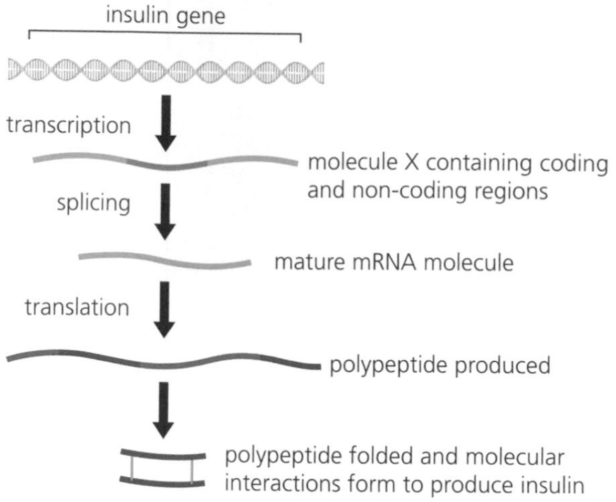

insulin gene

transcription → molecule X containing coding and non-coding regions

splicing → mature mRNA molecule

translation → polypeptide produced

polypeptide folded and molecular interactions form to produce insulin

a Name the coding regions of eukaryotic genes.

1 — Demonstrating KU

b Name molecule X which is transcribed from the DNA.

1 — Applying KU

c Explain the significance of alternative RNA splicing in terms of gene expression.

1 — Applying KU

d Name **one** molecular interaction which can produce the shape of the final protein such as in the case of insulin.

1 — Demonstrating KU

3 Severe combined immunodeficiency (SCID) is a rare inherited condition.

Gene therapy using bone marrow stem cells has been used in the treatment of some children with this condition.

a Give **two** properties of stem cells which make them suitable for use in this therapy.

1 _____

2 _____

2 — Applying KU

MARKS   STUDENT MARGIN

b The graph shows the number of functional white blood cells in a patient who has undergone this treatment. Unaffected children have a white blood cell count which ranges between 5000 and 8000 cells per mm³ of blood.

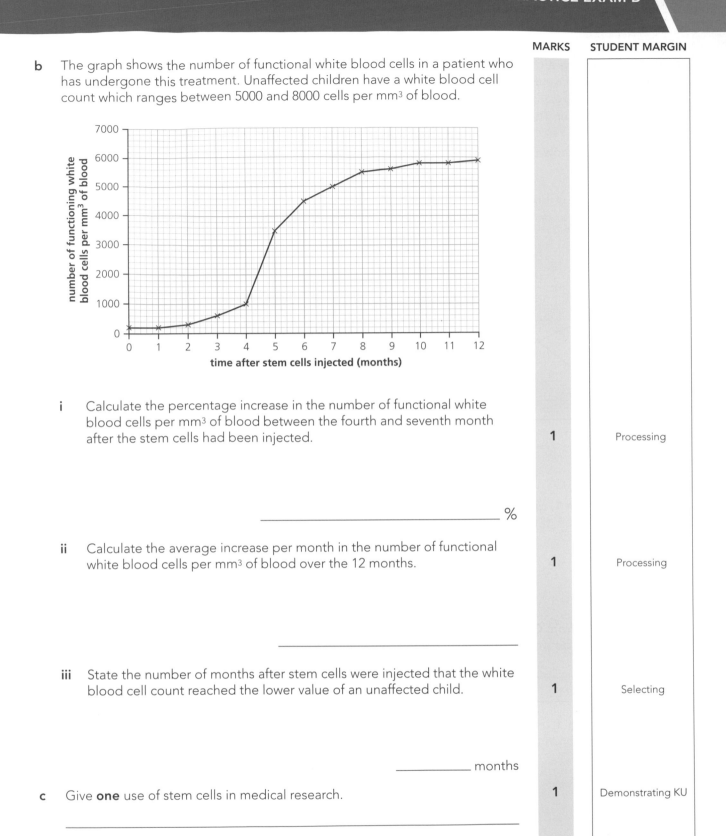

time after stem cells injected (months)

i Calculate the percentage increase in the number of functional white blood cells per mm³ of blood between the fourth and seventh month after the stem cells had been injected.

1   Processing

_____ %

ii Calculate the average increase per month in the number of functional white blood cells per mm³ of blood over the 12 months.

1   Processing

_____

iii State the number of months after stem cells were injected that the white blood cell count reached the lower value of an unaffected child.

1   Selecting

_____ months

c Give **one** use of stem cells in medical research.

1   Demonstrating KU

_____

_____

MARKS | STUDENT MARGIN

4   During pharmaceutical trials for a new drug, a healthy subject volunteered to drink one litre of water.

Samples of urine were taken from the subject over the 180-minute period after drinking the water. These samples were used to determine the rate of urine production. The salt concentration of the samples was also measured.

The results are shown in **Graph 1**.

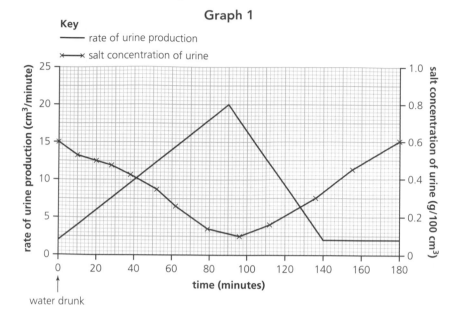

Graph 1

a   From **Graph 1**, give the time taken, after drinking the water, for the salt concentration of the urine to decrease to its lowest concentration.

1   Selecting

_____ minutes

b   Use values from **Graph 1** to describe the changes in the rate of urine production between 0 and 180 minutes.

2   Selecting

_____

_____

c   From **Graph 1**, identify the rate of urine production when the salt concentration in the urine was at its maximum over the 180-minute period.

1   Selecting

_____ cm³ per minute

MARKS | STUDENT MARGIN

d In a further investigation, 10 mg of the new drug under trial was administered to the volunteer who then drank a further one litre of water.

The rate of urine production was measured and the results are shown in **Graph 2**.

Graph 2

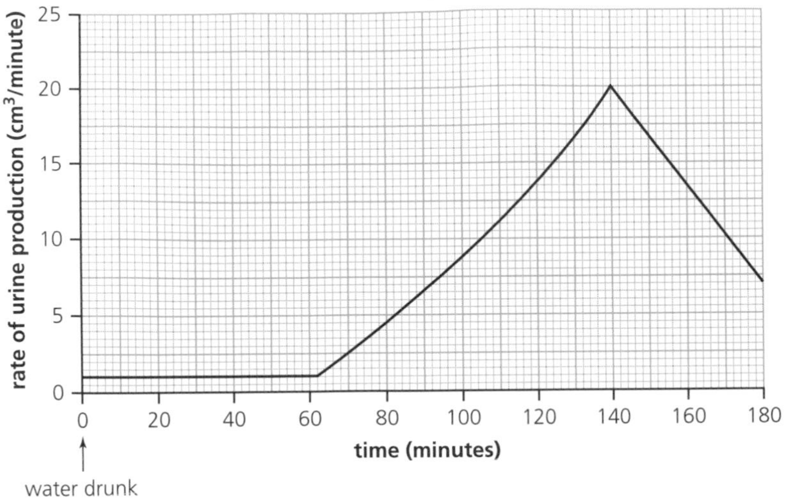

From **Graphs 1 and 2**, calculate the decrease in urine production due to the drug at 90 minutes into the trial.

1   Processing

_____ cm³

e The new drug has the potential to be used in the personalised medical treatment of individual patients.

Describe the type of analysis which would have to be undertaken so that the likelihood of success with the drug for a particular patient could be estimated.

1   Applying KU

_____

_____

5 Answer **either A or B** in the space below.

Labelled diagrams may be used where appropriate.

4   Demonstrating KU

A Give an account of the effects of stabilising selection on the variation of an animal population.

OR

B Give an account of the effects of directional selection on the variation of an animal population.

_____

_____

_____

_____

_____

_____

_____

_____

	MARKS	STUDENT MARGIN

6 The phylogenetic tree shows the evolutionary relationships between some animal species which exist today.

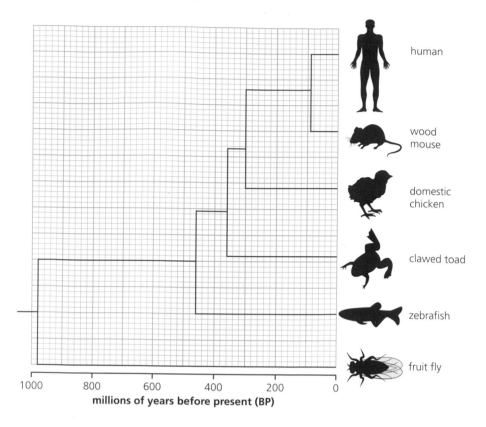

millions of years before present (BP)

a Give **two** sources of data which can be used to confirm the evolutionary relationships between the animal species shown in the diagram.

2   Demonstrating KU

_____

_____

b Calculate how many million years separated the divergence of fruit fly and other species in the diagram from the divergence of chickens and humans.

1   Processing

_____ million years

c State how long ago the last common ancestor of the wood mouse and the clawed toad existed.

1   Processing

_____ million years BP

d State the number of other animals in the diagram with which the zebrafish shared a common ancestor 450 million years before present.

1   Selecting

_____ species

| | MARKS | STUDENT MARGIN |

7 The diagram represents a section through the inner mitochondrial membrane and some of the processes of the electron transport chain leading to ATP formation.

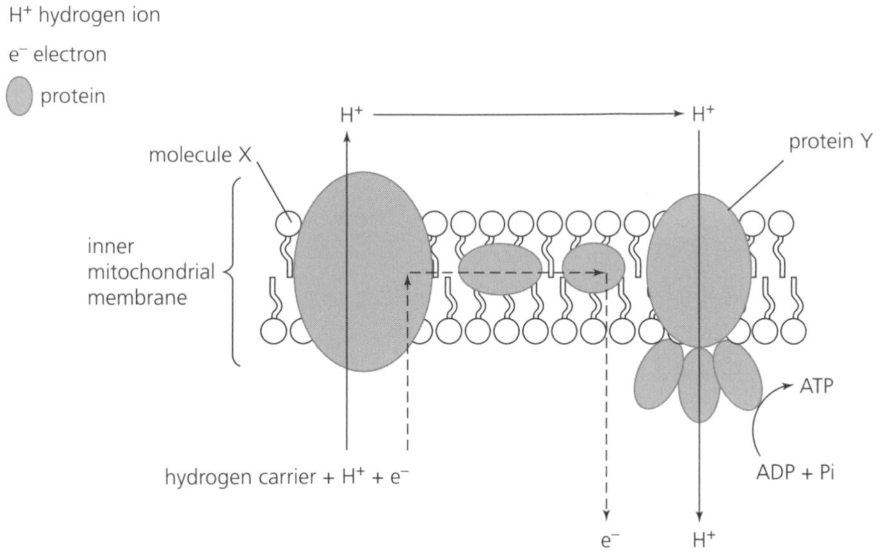

H+ hydrogen ion

e⁻ electron

⬭ protein

molecule X

inner mitochondrial membrane

H+ ⟶ H+

protein Y

ATP

hydrogen carrier + H+ + e⁻

ADP + Pi

e⁻   H+

a Name the coenzyme that carries hydrogen ions and electrons to the electron transport chain.

_____

1   Demonstrating KU

b Name molecule X which forms part of the inner mitochondrial membrane.

_____

1   Applying KU

c Describe the role of the electrons that pass down the electron transport chain.

_____

_____

1   Demonstrating KU

d Name protein Y, responsible for the regeneration of ATP.

_____

1   Applying KU

e Describe the role of oxygen in the electron transport chain.

_____

_____

2   Demonstrating KU

8 Catechol oxidase is an enzyme found in banana tissue. It is involved in a two-step reaction which produces the brown melanin pigment that forms in cut or damaged bananas.

The effect of the concentration of lead ethanoate on this reaction was investigated.

20 g of banana tissue was cut up, added to 20 cm³ of distilled water and then liquidised, filtered and a buffer added to keep the pH constant during the reaction. This produced a buffered extract containing both catechol and catechol oxidase.

Test tubes were set up as described in **Table 1** and kept at 20 °C in a water bath.

Every 10 minutes, the tubes were placed in a colorimeter which measured how much brown pigment was present.

The more brown pigment present, the higher the colorimeter reading. The results are shown in **Table 2**.

catechol ⟶ quinone ⟶ melanin pigment
(colourless substance   (yellow)   (brown)
in banana tissue)

Table 1

Test tube	Contents of test tubes
A	sample of extract + 1 cm³ 0.01% lead ethanoate solution
B	sample of extract + 1 cm³ 0.1% lead ethanoate solution

Table 2

Time (minutes)	Colorimeter reading (units)	
	Test tube A	Test tube B
0	1.8	1.6
10	5.0	2.0
20	6.0	2.2
30	6.4	2.4
40	7.0	2.4
50	7.6	2.4
60	7.6	2.4

a   Identify **two** variables, not already mentioned, that would have to be kept constant.

_____

_____

2   Planning

b   Explain why the initial colorimeter readings were not 0.0 units

_____

_____

1   Applying KU

c   Describe a suitable control for this investigation.

_____

_____

1   Planning

	MARKS	STUDENT MARGIN

d The results for the extract with 0.1% lead ethanoate are shown in the graph.

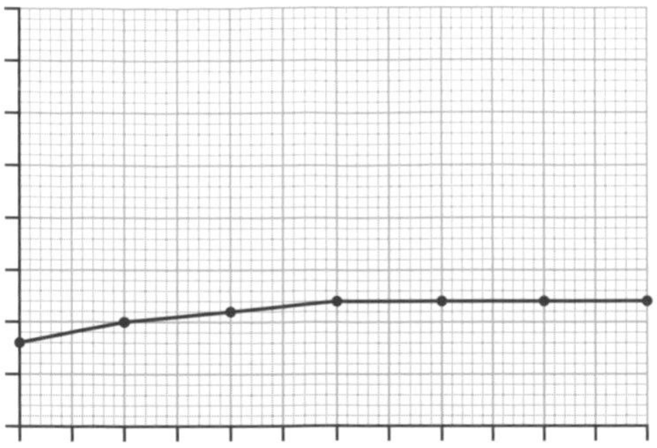
extract + 0.1%
lead ethanoate

Use information from **Table 2** to complete the graph by:

i adding the scale and label to each axis — 1 — Presenting

ii presenting the results for the extract + 0.01% lead ethanoate solution and labelling the line. — 1 — Presenting

e Give a conclusion which can be drawn about the effect of the concentration of lead ethanoate solution on the activity of catechol oxidase. — 1 — Concluding

____

____

f State how the procedure could be improved to increase the reliability of the results. — 1 — Evaluating

____

____

9 The diagram represents a vertical section through a part of the human skin.

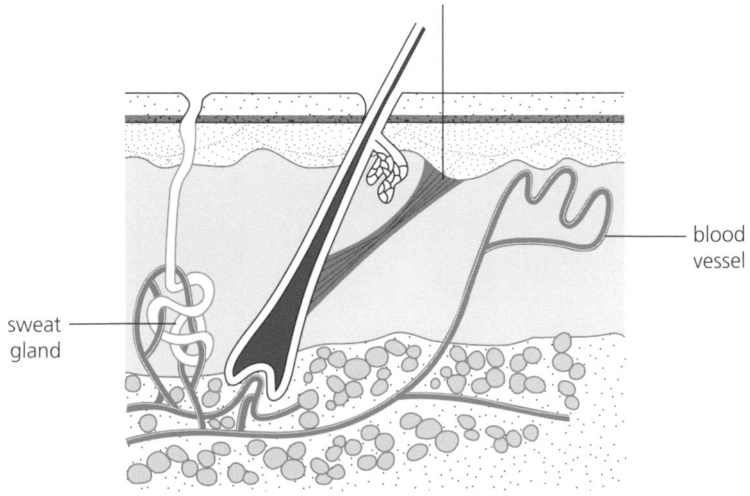
hair erector muscle
blood vessel
sweat gland

a Select **one** structure labelled in the diagram and explain its role in response to a **decrease** in temperature. — 1 — Applying KU

Structure: ______________________

____

____

MARKS | STUDENT MARGIN

b Explain the importance of regulating body temperature to the metabolism of humans.

2 | Applying KU

_____

_____

c Give the term used to describe organisms whose internal environment is dependent upon their external environment.

1 | Demonstrating KU

_____

10 Plasmids are often used as vectors in recombinant DNA technology.

Many different restriction endonuclease enzymes are used in the process and each enzyme cuts the DNA of the plasmid at a specific base sequence called its restriction site.

The diagram shows the position of four different restriction sites: J, K, L and M. The distances between the restriction sites are measured in kilobases (kb) of DNA as shown.

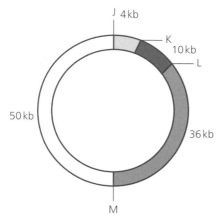

Two restriction endonuclease enzymes were used to cut the plasmid DNA. The fragments of plasmid DNA which resulted were separated by gel electrophoresis.

The positions of the fragments in the electrophoresis gel are shown in the **chart**.

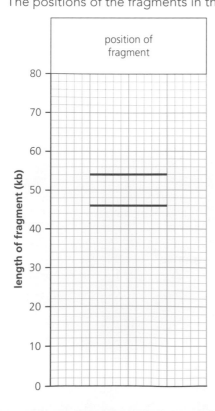

		MARKS	STUDENT MARGIN

a **i** From the information in the **chart** and **diagram**, state which restriction sites were cut.

     Site _____ and site _____

MARKS: 1 — Selecting

  ii Predict the length of the DNA fragment expected if the plasmid had been cut at one restriction site only.

                                               _____ kb

MARKS: 1 — Predicting

b Name the enzyme which can be used to seal the foreign DNA into the plasmid.

  _____

MARKS: 1 — Demonstrating KU

c State **one** disadvantage of using bacterial cells, such as *E. coli*, as the recipient for foreign DNA.

  _____

  _____

MARKS: 1 — Demonstrating KU

d Recombinant DNA technology is one method of producing improved varieties of bacteria.

  Name **one** other method of producing an improved variety of bacteria **and** describe how it can be carried out.

  Name: _____

  Description: _____

  _____

MARKS: 1 — Demonstrating KU

11 The graph shows the growth of a population of a bacterial culture added to a sterile fermenter over a 60-hour period.

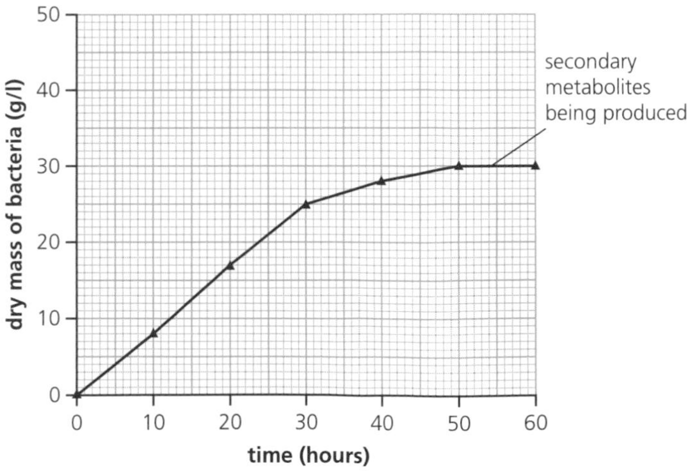

a **i** Explain why the fermenter was sterilised before the culture of bacteria was added.

    _____

    _____

MARKS: 1 — Demonstrating KU

  ii State **one** other condition within the fermenter that would be controlled to provide optimum culture conditions for growth of the bacteria and describe how this would be achieved.

    Condition: _____

    Description: _____

    _____

MARKS: 2 — Demonstrating KU

	MARKS	STUDENT MARGIN

iii Calculate the average hourly growth rate of the bacteria over the 60-hour period.

1   Processing

_____ per hour

iv Predict what will happen to the dry mass of live bacteria after 60 hours **and** justify your answer.

1   Processing

_____

_____

b Secondary metabolites are not associated with growth and can confer ecological advantage to the bacteria which produce them.

Give **one** example of a secondary metabolite and explain how it can confer an ecological advantage.

2   Demonstrating KU

_____

_____

12 Some birds which breed in Britain migrate south in winter. Some species winter in temperate regions south of Britain but others travel further to tropical areas.

The graph shows how the migration rate of tropical and temperate wintering species varies with the time they depart their breeding areas.

Key
— species which winter in tropical areas
- - - species which winter in temperate areas

a **i** Describe the relationship between departure time and the migration rate for species which winter in tropical areas.

1   Selecting

_____

_____

ii Describe how departure dates and migration rates compare between species which winter in tropical areas and those which winter in temperate areas.

1   Concluding

_____

_____

iii Suggest why the species which winter in the tropics generally depart earlier than those which winter in temperate regions.

1   Selecting

_____

_____

	MARKS	STUDENT MARGIN

b i Describe **one** technique which could be used to collect data on migration rates of birds.

_____

1 — Demonstrating KU

ii Migration is an adaptation to avoid adverse conditions.

Name **one** other adaptation used by some species to tolerate adverse conditions.

_____

1 — Demonstrating KU

13 Fire ants *Solenopsis invicta* were accidentally brought to the USA by ships from South America during the 1930s. The imported fire ants spread quickly and displaced native ant species.

a Fire ants are now classified as an invasive species in the USA.

Give **one** reason why the population of an invasive species may increase at the expense of native species.

_____

_____

1 — Demonstrating KU

b Scuttle flies attack fire ants in South America. Female scuttle flies lay eggs in the ant's body which hatch to larvae that feed on its internal tissues.

An investigation was carried out into the control of fire ants using an integrated pest management (IPM) approach which combined the use of scuttle flies and an insecticide.

The graph shows the results.

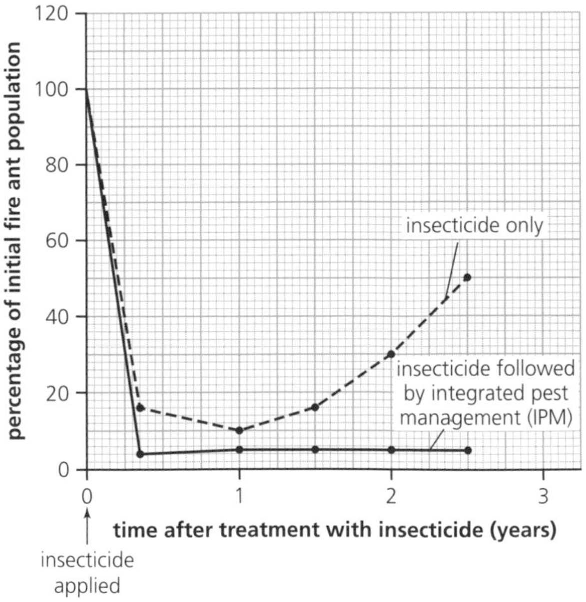

i Compare the effect of using the insecticide only with the integrated approach using insecticide and scuttle flies.

_____

_____

2 — Concluding

ii Suggest how natural selection might account for the rise in the fire ant population which occurred with the insecticide-only treatment after year 1.

_____

_____

2 — Demonstrating KU

	MARKS	STUDENT MARGIN

iii Describe **one** possible risk of introducing scuttle flies to control fire ant populations in the USA.

_____

_____

1 — Applying KU

iv Explain how the information given about the scuttle flies confirms that they have a parasitic relationship with the fire ants.

_____

_____

1 — Applying KU

14 Pigs are sometimes reared in intensive units in which temperature is controlled. The effect of temperature on growth rate and the efficiency with which the pigs converted food to biomass was investigated and the results are shown in the table.

Temperature (°C)	Mean growth rate (g/day)	Conversion efficiency of food to biomass
0	540	19
10	800	42
20	850	48
30	450	37
35	310	37

a Explain why pigs of the same breed, with similar genotypes, were used in this investigation.

_____

_____

1 — Applying KU

b i Describe the effect of temperature on mean growth rate of the pigs.

_____

_____

2 — Selecting

ii Express the mean growth rate at 30 °C to 0 °C as the simplest whole number ratio.

1 — Processing

_____ : _____
at 30 °C at 0 °C

iii Predict the conversion efficiency of food to biomass by pigs kept at a temperature of 15 °C.

1 — Predicting

_____

c Growing crops for human food is a more efficient use of resources than the rearing of stock animals like pigs.

Suggest an explanation for this.

_____

_____

2 — Applying KU

d Give **one** example of a type of behaviour which could indicate poor welfare in domesticated pigs.

_____

1 — Demonstrating KU

	MARKS	STUDENT MARGIN

15 The diagram shows crosses in a breeding programme involving different breeds of pig. The parent (P) breeds are produced using inbreeding techniques.

Parent (P)   Hampshire × Yorkshire

F_1 Blue butt × Duroc

F_2 Commercial piglets for market

a Suggest why the parent breeds have been produced using inbreeding.

_____

_____

	MARKS	STUDENT MARGIN
a	1	Demonstrating KU

b Suggest a reason why the inbred parents are crossbred in programmes such as this.

_____

_____

	1	Applying KU

c Explain why Blue butt pigs are produced by crossbreeding Hampshire and Yorkshire pigs instead of breeding F_1 Blue butts together.

_____

_____

	1	Applying KU

d The Duroc males selected for the F_1 cross are homozygous for desired dominant characteristics.
Explain the importance of this selection.

_____

_____

	1	Demonstrating KU

16 Answer either **A or B** in the space below.
Labelled diagrams may be used as appropriate.

A Write notes on social behaviour in animals under the following headings:

   i social hierarchy

   ii cooperative hunting and social defence.

	MARKS	STUDENT MARGIN
i	4	Demonstrating KU
ii	5	Demonstrating KU

OR

B Give an account of photosynthesis under the following headings:

   i the capture and use of light energy

   ii the Calvin cycle.

	MARKS	STUDENT MARGIN
i	5	Demonstrating KU
ii	4	Demonstrating KU

_____

_____

_____

_____

_____

_____

_____

_____

_____

_____

_____

[End of Practice Exam B]

Practice Exam A

Paper 1

Multiple-choice				Commentary with hints and tips
Question	Answer	Mark	Demand	
1	C	1	C	The answer here is about applying the complementary base pairing rules of **A**denine with **T**hymine and **G**uanine with **C**ytosine. You need to remember that the two daughter DNA molecules are identical.
2	D	1	C	Base pairing rules are needed again but remember that **U**racil is not found in DNA. At the end of a DNA molecule, each strand has a different prime – a 3' on one strand and a 5' on the other.
3	A	1	C	Prokaryotes don't have any membrane bound organelles, therefore they have no nucleus, mitochondria or chloroplasts but, like eukaryotes, they do synthesise proteins so ribosomes are present.
4	A	1	C	Remember **SID** – the gene mutations are **S**ubstitution, **I**nsertion and **D**eletion. The other names given refer to chromosome mutations – watch out though, there is also a chromosome mutation called Deletion!
5	C	1	C	When DNA is transcribed, RNA is produced; if the DNA is coding, mRNA is produced, but if it is non-coding, either rRNA or tRNA is produced by transcription.
6	A	1	A	Work out 1% of 6×10^9 which is 6×10^7, then multiply by 4 to find 4% which is 2.4×10^8, then divide by 3 to convert bases to codons which gives the answer, 8×10^7.
7	B	1	C	The mean wing length has increased so selection pressure is directional and acts against shorter wings.
8	D	1	C	This question requires checking the time of the last common ancestors of the groups – the longer ago the common ancestor existed the less closely related the groups.
9	D	1	C	The active site shape is complementary to the substrate to which it is bound so we know that I must bind to another site which means it must be non-competitive. Then we must remember that non-competitive inhibition is not affected by substrate concentration.
10	C	1	A	The longer it takes for complete breakdown, the slower the enzyme is working so you must compare the values at high concentration with the values when 0 ions were present. In each case, the reaction takes longer.
11	B	1	A	Ethanol is made by breakdown of pyruvate so if a yeast cell can't make pyruvate or break it down, ethanol production is not possible; citrate is made from pyruvate so if citrate is present there must be pyruvate to make it from – this is tricky!
12	D	1	C	Remember **PEP**: **P**umps, **E**nzymes and **P**ores – the functions of membrane-bound proteins. A great example is to remember the electron transport chain which has all three types working together.
13	D	1	C	Birds have a high metabolic rate so need to deliver oxygenated blood to cells with complete separation from deoxygenated blood which in turn requires two ventricles and two atria.
14	C	1	C	The greater the metabolic rate, the greater the consumption of oxygen, so you need to look at the points on the graph where the metabolic rates are highest – this is quite straightforward.

Multiple-choice				Commentary with hints and tips
Question	Answer	Mark	Demand	
15	B	1	A	You are looking for procedures which can change the DNA sequences of the bacteria – in other words, cause mutations. Remember that radiation and certain chemicals can do this.
16	C	1	A	Plasmids are tiny, so fitting in large sections of genetic information is not possible and artificial chromosomes are created to do this.
17	D	1	C	Remember that chlorophyll absorbs in the blue and red regions of the spectrum – only D corresponds with that pattern.
18	B	1	C	Carotenoids are the pigments which capture wavelengths of light not absorbed by chlorophyll; they pass their absorbed energy to chlorophyll to allow photosynthesis. They do not take energy from chlorophyll.
19	D	1	C	Chimpanzees are likely to feed preferentially, eating the foods they like best first just as they would do naturally. All hens should preen their feathers – it's over-preening which could indicate poor welfare.
20	B	1	C	Introduced species have been moved by humans to new geographical locations whereas indigenous species are native to a region. Invasive species become naturalised then go on to damage natural communities in various ways.
21	A	1	A	Crossbreeding combines the characteristics of two true breeds and increases heterozygosity and vigour. Remember, crossbreeds are not bred on because their offspring will be too varied.
22	A	1	C	Annual weeds grow from seed and complete their life cycle each year; they don't overwinter as plants so don't need storage organs for winter food.
23	A	1	C	Energy is lost at each trophic level so the fewer trophic levels involved the more energy is retained; it's better to eat the plant crop than to feed it to animals which use 90% of the energy in their food to survive.
24	C	1	C	Really just a matter of reading the options one by one and eliminating those which are clearly wrong when applied and compared with the graph values.
25	D	1	A	Tricky – you really have to read values in each of the years in the possible answers. Notice that the scales on the two y-axis are very different in magnitude.

Paper 2

Question			Expected answer	Mark	Demand	Commentary with hints and tips
1	a		RNA is single-stranded **AND** DNA is double-stranded **OR** RNA is non-helical **AND** DNA is helical **OR** RNA has ribose **AND** DNA has deoxyribose **OR** RNA has uracil **AND** DNA has thymine **(Any 2, 1 mark each)**	2	CC	RNA molecules are differently shaped to DNA molecules and their nucleotides have different sugars; in RNA the base uracil replaces thymine as complementary to adenine. Remember, this question referred to structural differences but you could be asked about functional differences too, so make sure you know them!
	b		RNA polymerase adds complementary RNA nucleotides to form a primary RNA transcript = **1** Unwinds and unzips the double helix of DNA = **1**	2	CA	Remember to refer to the **complementary** nucleotides as RNA nucleotides; remember that the enzyme also unwinds and unzips the DNA.
	c		Introns are removed from the primary transcript = **1** Exons are joined/spliced together to make mature mRNA = **1**	2	CC	Introns are always removed; in alternative splicing some exons are treated as introns and removed to produce the mature mRNA. Remember – **ex**ons are **ex**pressed and **in**trons **in**terrupt the gene.
	d		Carries a complementary copy of the DNA code from the nucleus to the cytoplasm/ribosome	1	C	It's good practice to say that the copy is complementary and state where the copy is coming from and where it is going to.
2	a	i	C G U	1	C	Just have to remember to use Uracil not Thymine here.
		ii	Label at longer end of the RNA chain on the left of the diagram	1	C	The amino acid is bonded to the 3' end of the RNA chain which sticks out beyond the 5' end – you don't need this detail, only the place it bonds.
	b		An anticodon is a group of three bases on the tRNA (which relates to a specific amino acid) = **1** The anticodon aligns with/is complementary to a codon on mRNA to bring its specific amino acid into place (on the forming polypeptide) = **1**	2	CA	Tricky to put into words but the key words are specific, complementary and aligns. Try to use all of these words in this sort of description.

Question		Expected answer	Mark	Demand	Commentary with hints and tips
	c	Transcribed to rRNA **OR** regulatory sequences	1	C	rRNA combines with protein to form ribosomes; regulatory sequences can affect the expression of a gene by controlling transcription including turning it off or on.
3	a	Involves geographical isolation between populations = **1** which acts as a barrier to gene/mutation flow prevents interbreeding between populations = **1** After a long period of time, genetic differences build up on either side of the barrier = **1**	3	CCA	Again, the key words make scoring marks easier – try to use isolation, geographical barrier and gene flow; remember, a long time period of isolation is needed for a new species to form.
	b	The offspring produced by interbreeding in area A would be sterile/less fit	1	C	Sterile means that the hybrids don't produce their own offspring. This confirms that the parents were of two different species.
4	a	A B d e f	1	C	Remember to read off the new allele sequence following the process – just as has been done for gamete 3 in the table already.
	b	Name: Duplication = **1** Importance: Allows gene mutation of duplicated region = **1** Without losing the function of the original gene = **1**	3	CAA	Tricky – the duplicated gene could be mutated and possibly produce an advantage but the function of the original gene is preserved.
5	a	Same contents as the experimental tube but without the DNA sample	1	A	If any DNA is amplified in this control, it has to come from contamination.
	b	The polymerase used is heat tolerant (from bacteria which live in hot springs)	1	C	The high temperatures are used to separate the strands of the sample DNA.
	c	To maintain a steady pH **OR** To resist pH changes	1	C	Remember the role of buffers in fermenters.
	d	To help solve crimes **OR** To settle paternity suits **OR** To diagnose genetic disorders	1	C	List straight from the course specifications.

	Question		Expected answer	Mark	Demand	Commentary with hints and tips
6	a		Axes, scales and labels with units = **1** Points plotted accurately and connected with straight lines = **1**	2	CC	Ensure that the origin point is coordinated – use two zeros if possible; divide scales evenly; take labels with units directly from the data table in the question. Plot points as dots and connect using a ruler.
	b	i	Time for methylene blue to decolourise **OR** Rate of respiration	1	C	You need to know the difference between the dependent variable which is the result of the experiment and the independent variable which causes the result in a controlled experiment.
		ii	Concentration of yeast/glucose/methylene blue pH species of yeast **(Any 1)**	1	C	Remember that all variables must be kept the same apart from the independent variable which is under investigation.
	c	i	Set up a tube in exactly the same way but replace the yeast with distilled water/leave out the yeast/use killed yeast	1	C	Controls differ from experiments in only one way and can be compared with the experiment to ensure that the independent variable was causing the result.
		ii	To ensure that the contents of the tubes were all at the correct temperature	1	A	If the contents were mixed without bringing them to the required temperature, respiration would start at the mixing temperature not the experimental temperature wanted.
	d		Higher temperature has denatured the enzymes involved in respiration (and reduced the rate of respiration)	1	A	It's important to link temperature with the rate of enzyme reactions.
7	a	i	125%	1	A	Find the difference between the end August value and the end April value = 2.0, then divide this by the August value = 1.6, then multiply the answer by 100. 2.0 ÷ 1.6 × 100 = 125%
		ii	2.0 mg per 100 g dry weight	1	C	Simply read the values from the graph and find the difference. 2.5 − 0.5 = 2.0
	b		The maximum level of sucrose in the needles in winter varies by the end of each winter = 1 Winter temperatures expected to vary = 1	2	CA	This answer requires looking at the graph for *P. strobus* closely – the winter levels of sucrose vary which fits with the tree responding to varied winter temperatures.

Question			Expected answer	Mark	Demand	Commentary with hints and tips
	c		Winter temperatures are higher in the range of *P. virginiana* than in the range of *P. strobus* = **1** Levels of sucrose are much higher in *P. strobus* = **1**	2	CA	The question stem suggests that the adaptation to cold is increased sucrose levels. *P. virginiana* has overall lower sucrose levels, suggesting it is not so highly adapted to cold.
	d		Migrate **OR** Relocate to another area	1	C	Notice that the question asks for **avoidance** and not toleration – migration is really the only way.
8	a	i	Blood passes through the heart only once for a complete circulation of the body	1	C	Fish are the only vertebrates with single circulation and a simple two-chambered heart.
		ii	Double circulation allows blood pressure to be maintained at higher levels (because the blood passes through the heart twice per circulation) **OR** Complete double circulatory systems enable higher metabolic rates to be maintained **OR** The oxygenated blood can be pumped out at a higher pressure **OR** It enables more efficient oxygen delivery to cells	1	A	Blood pressure is vital to ensure efficient oxygen delivery to cells and sending the blood twice through the heart for each circulation is an adaptation to achieve this.
	b		In vessel X blood is entering the heart and in vessel Y blood is leaving the heart	1	A	Definitely worth learning the direction of blood flow through hearts – blood always enters via atria and leaves via ventricles.
	c		Structure Z separates oxygenated and deoxygenated blood = **1** so pure oxygenated blood passes to body tissues = **1**	2	CA	The separation into two ventricles is what makes bird and mammal hearts so efficient at keeping blood with oxygen from mixing with blood without.
9	a		Protein	1	C	All membranes have proteins.
	b		The flow of electrons along the transport chains provides the energy to pump the ions (against their concentration gradient) into the space	1	A	The source of the energy is the key point here.
	c		Hydrogen ions flow down their concentration gradient = **1** Through ATP synthase = **1**	2	CA	ATP is generated from ADP and Pi as the ions flow through.

	Question		Expected answer	Mark	Demand	Commentary with hints and tips
10	a	i	Remove any competing microorganisms/pathogens	1	C	Remember **stop** for the conditions inside fermenters – **s**terility, **t**emperature control, **o**xygen control and **p**H control.
		ii	To mix the contents to prevent the culture settling **OR** To help aerate/oxygenate the culture	1	A	The cells in the culture are denser than the medium so gravity will cause them to settle. Mixing also helps to supply oxygen to the cells for aerobic respiration.
	b		To keep them at optimum levels for enzyme action	1	C	Temperature and pH are factors which affect enzymes and enzymes are needed for respiration; using the word optimum here will help you to gain the mark.
	c		Glucose = **1**; for energy = **1** **OR** Amino acids/vitamins/fatty acids = **1**; for biosynthesis/ growth = **1**	2	CA	The nutrients must supply energy to the bacteria and although some species can make all the other substances they need for biosynthesis, some have to have these supplied.
11	a	i	Ligase can be used to seal the sticky ends	1	A	All ligases join molecules together – a similar ligase is used to join fragments of DNA during replication of lagging strands; remember that it's sticky ends which are being joined.
		ii	Can be used to identify those bacteria which have taken up the marker and so also the desired gene	1	C	Adding plasmids to bacterial cells is tricky and has a failure rate. Marker genes help to identify cells which have successfully taken up the plasmid.
	b		Yeast are eukaryotes and so can fold polypeptides correctly **OR** Bacteria do not always fold polypeptides correctly	1	C	Only eukaryotes can fold polypeptides so to produce a functional eukaryote protein, a eukaryotic cell is needed; yeast is a special eukaryote.
	c		Horizontal gene transfer	1	C	In horizontal gene transfer, DNA sequences can be passed between members of one generation.
12	a	i	Mass of herbicide applied increased from 7000 tonnes in 1970 to 10 500 tonnes in 1990 = **1** Then decreased from 10 500 tonnes in 1990 to 6000 tonnes in 2000 = **1**	2	CC	Remember to quote values with units from each axis and to look for trends rather than details.
		ii	50%	1	C	Rise of 3500 tonnes between 1970 and 1990, so 3500 ÷ 7000 × 100%

Question			Expected answer	Mark	Demand	Commentary with hints and tips
		iii	0.175 million/175 000 hectares per year	1	A	5.25 million increase over 30 years so 5.25 million ÷ 30 = 0.175 million or 175 000 hectares.
		iv	1312.5 tonnes	1	A	Read the data for 1990; 10 500 tonnes ÷ 8
	b	i	1990	1	A	Tricky – note it is a 40% reduction which would take the population down to 60% of what it was. You need to read the scale at 60% not 40%.
		ii	24%	1	A	Simply extend the line down until it lines up above 2002.
	c		Herbicide has killed more weeds so there is less food for birds **OR** Herbicide could be poisoning/ killing/affecting the reproduction of the birds/impacting the birds directly	1	C	Have to infer an answer here but food is the main suggestion.
	d		Mass of herbicide applied dropped between 1990 and 2000 (even although the area treated increased)	1	C	In recent years, farmers have got better at targeting their crops using less herbicide but better placed.
13	A		**1** Genetic diversity **2** Genetic diversity is the number and frequency of alleles in a population **3** Species diversity **4** Species diversity is species richness **AND** relative abundance **5** Ecosystem diversity **6** Ecosystem diversity is the number of different/distinct ecosystems (in a defined area) **(Any 4, 1 mark each)**	4	CCCA	The answer relies on remembering that there are three categories of biodiversity.
	B		**1** Invasive species are naturalised species that spread rapidly **2** And can reduce biodiversity **3** Invasive species have no natural predators **4** Invasive species have no natural parasites/diseases **5** Can eliminate native species by outcompeting them **6** Affect native species by preying on them **7** Affect native species by hybridising with them **(Any 4, 1 mark each)**	4	CCCA	This answer relies on remembering what invasive species are and how they affect native species.

Question			Expected answer	Mark	Demand	Commentary with hints and tips
14	a		Chemical = **1** Use of a natural predator/parasite/disease of pest = **1** Crop rotation/ploughing/weeding = **1**	3	CCA	It is useful to remember the three pest management categories and an example of each one.
	b		Persistent in the environment **OR** Toxic to (non-pest) animal species **OR** Accumulate/be magnified in food chains **OR** Produce resistant populations	1	C	Although pesticides have been very useful in increasing crop yields, there are environmental impacts to take into account.
15	a	i	3-phosphoglycerate, aspartic acid, glycine, alanine	1	C	New spots on the chromatogram show the substances which have appeared since the previous chromatogram.
		ii	0.8	1	C	Looking across the diagram shows that alanine has moved 8 units but the solvent front has moved 10 units. $8 \div 10 = 0.8$
	b	i	In light conditions, 3-phosphoglycerate decreases as RuBP increases	1	A	As a substrate is converted to a product in a metabolic pathway, its concentration decreases and that of the product increases – this is what the graph shows.
		ii	3-phosphoglycerate/3PG = **1** Photolysis = **1**	2	CC	3-phosphoglycerate is produced after RuBP accepts CO_2 from the air in carbon fixation – it is 3-phosphoglycerate which is reduced by hydrogen from the splitting of water by photolysis to form glyceraldehyde-3-phosphate (G3P).
16	a		(It is a co-evolved relationship) in which the two different species involved both benefit	1	C	The emphasis here is one benefit to **both species** – don't say both animals or just both.
	b	i	Parasite species benefits in terms of energy or nutrients **AND** host is harmed by loss of energy or nutrients	1	A	Concentrate on what happens to energy and nutrients when talking about parasitism.
		ii	Parasites (often) have limited metabolism	1	C	Limited metabolism is the key here – don't write about nutrients or energy.
		iii	Direct contact **OR** Vectors **OR** Resistant stages	1	C	Don't try to describe these, just give the names.

	Question		Expected answer	Mark	Demand	Commentary with hints and tips
17	a	i	50%	1	A	The chimp is 2.8 years old and from the table 1.4 years have been spent travelling on its mother's back: $1.4 \div 2.8 \times 100\%$
		ii	The 2.9-year-old would be within 5 metres of its mother but the 3.6-year-old could be further than 5 metres away	1	C	Reading the ages from the table shows the 5-metre difference.
	b	i	Time to learn complex social skills	1	C	The same as for any primate including humans.
		ii	Grooming Facial expression Body posture Sexual presentation Threat display Appeasement display Ritualistic display (Any 2, 1 mark each)	2	CC	These are the behaviours mentioned in the course specification so safest to score marks.
18	A	i	1 Unspecialised/undifferentiated cells in animals 2 Can divide to form new stem cells 3 Differentiate into specialised cells 4 Which develop more specialised functions 5 By expressing genes characteristic of that cell type 6 Embryonic stem cells differentiate into almost all cell types/are pluripotent 7 Adult (tissue) stem cells differentiate only into cells of the tissue type in which they are found/are multipotent (Any 5, 1 mark each)	5	CCCCA	Remember, stem cells do TWO things – they can divide AND some differentiate to allow growth and repair in animals.
		ii	Research provides information about: 1 how cell growth/differentiation works 2 how gene regulation works Can be used as model cells: 3 to study how diseases develop 4 for testing new drugs 5 Therapeutic use in replacing damaged or diseased organs/tissues **OR** example (Any 3, 1 mark each)	3	CCA	Research is about studying and learning but therapy is about treating and curing.

Question			Expected answer	Mark	Demand	Commentary with hints and tips
	B	i	**1** Genomic sequences are the sequences of nucleotide bases in DNA **2** Studying genomic sequence data and fossil records can determine the order of evolutionary events **3** Comparison of genomic sequences requires bioinformatics **4** Sequence comparison provides evidence for three domains of life **OR** for bacteria and archaea and eukaryotes **5** The main events in evolution are last common ancestor/prokaryotes/photosynthesis/eukaryotes/multicellularity **(Any 3, 1 mark each)**	3	CCA	Genomic sequence data can show the relationships between species – how similar they are and how long ago their last common ancestor existed.
		ii	**1** Personal genomics involves getting sequence data for an individual **2** Analysis of an individual's genome may lead to personalised medicine **3** Pharmacogenetics is use of drugs according to an individual's genome **4** Identification of genetic components of disease **5** Increased likelihood of success of an individual treatment **6** Difficulties with personalised medicine include presence of neutral mutations **7** And the estimating of the genetic components of a specific disease **(Any 5, 1 mark each)**	5	CCCCA	Individuals of a species have slight differences in their genomic sequences which could form the basis of personalising a person's treatment if they become ill.

Practice Exam B
Paper 1

Multiple-choice				
Question	Answer	Mark	Demand	Commentary with hints and tips
1	C	1	C	Eukaryotic cells including animals, plants and fungi such as yeast contain linear chromosomes in their nucleus but also have circular chromosomes in their mitochondria and in the chloroplasts of plants.
2	D	1	C	80 base pairs means there are 160 nucleotides and so 160 deoxyribose molecules.
3	B	1	C	Remember: meristems contain undifferentiated cells, embryonic cells are pluripotent and can differentiate into all cell types and adult stem cells are multipotent and give rise to a more limited range of cell types.
4	A	1	C	Coding sequences in the genome are transcribed and translated so mRNA is needed.
5	B	1	A	Highest glutaminase activity is on day 2. The stem cell number on the same day is 3.
6	D	1	C	Frameshift effects are seen when the number of nucleotides in the sequence is changed so adding by insertion and removing by deletion produce these effects.
7	D	1	A	A is tempting because mutations are random, but that does not affect assumptions made by molecular clocks – they assume constant rates of mutation like the ticks of a real clock.
8	A	1	C	Anabolic = building up and using energy. Catabolic = breaking down and releasing energy.
9	D	1	A	Tricky – have to realise that if there is a lower recapture rate, this means the moths are likely to have been predated which suggests they were at a selective disadvantage.
10	C	1	C	You need to know this process and the enzymes involved.
11	D	1	A	At 2% oxygen, the glucose breakdown is 25 units **per minute**. This is then multiplied by 5 to give 125. Watch out for this type of question.
12	C	1	C	Feedback inhibition occurs when an end product inhibits an enzyme **earlier** in the pathway.
13	B	1	C	Responding to the condition when it arises is consequential. Aestas = summer. Hibernus = winter.
14	D	1	A	3.5 mm = 3500 micrometers; 3500 ÷ 700 = 5.0
15	A	1	A	Chambers receiving blood are atria and pressure reduces as blood passes further from the heart.
16	D	1	C	'Action' implies the photosynthesis activity not just absorbing light.
17	B	1	C	Photolysis – the splitting of water – occurs in the light dependent stage of photosynthesis.
18	A	1	A	Natural selection removes any individuals that become homozygous for deleterious (harmful) recessive alleles.
19	B	1	C	All these terms seem plausible so learning the definitions is important – using flash cards is recommended.
20	D	1	C	Specific pest species die out if their preferred crop is not available each year – crop rotation is a method of ensuring this. Cultural methods involve human behavior such as rotating crops.

Multiple-choice				
Question	Answer	Mark	Demand	Commentary with hints and tips
21	C	1	C	Remember that in reciprocal altruism, the altruistic behaviour roles are later reversed.
22	B	1	C	In mutualism both species benefit from the relationship.
23	C	1	C	Remember, workers cooperate to raise the offspring of close relatives because they share genes.
24	B	1	A	Tricky – take all algebraic values from the table: $50 \div (5 \times 4) = 50 \div 20 = 2.5$
25	D	1	C	The smallest and most remote fragments have the lowest biodiversity.

Paper 2

Question			Expected answer	Mark	Demand	Commentary with hints and tips
1	a	i	Phosphate shown on the 5' end of the strand	1	C	DNA polymerase replicates DNA from the 3' end of the lead strand; the lagging strand has 5' on its end as shown.
		ii	Adenine	1	C	The base it is pairing with must be thymine because that base used to pair with adenine on the original stand as labelled in the diagram.
		iii	Nucleotide	1	C	Basic unit of DNA structure – a sugar, a phosphate and a base.
	b	i	90 °C	1	C	Remember **PCR '967'**: 90 °C separates DNA strands 60 °C allows primers to attach 70 °C optimum temperature for DNA polymerase
		ii	32	1	C	$1 \to 2 \to 4 \to 8 \to 16 \to 32$
		iii	Gel electrophoresis	1	C	One of the six techniques you need to know for your exam.
		iv	To allow an individual to be placed at the scene of a crime/solving crimes/forensics **OR** Paternity/maternity tests	1	C	Probably the best answers but others may be acceptable.
2	a		Exons	1	C	Remember – **ex**ons are **ex**pressed in gene **ex**pression, **in**trons **in**terrupt the gene sequence so are removed.
	b		Primary mRNA transcript	1	C	Primary mRNA contains introns that need to be removed.

Question			Expected answer	Mark	Demand	Commentary with hints and tips
	c		Different proteins can be expressed from one gene	1	A	In this process, particular exons of a gene may be included within or excluded from the final, processed messenger RNA (mRNA) produced from that gene.
	d		Hydrogen bonds	1	C	Our 20 000 genes can produce 1 million different proteins. This is achieved by both alternative RNA splicing and folding of the polypeptide chains.
3	a		**1** Unspecialised cells **2** Able to divide **OR** Able to differentiate into other cell types	2	CA	Unspecialised cells are cells which have not differentiated to perform a specific function. Stem cells divide by mitosis to produce more stem cells. Specialised cells express the genes characteristic of that cell type.
	b	i	400%	1	C	Increase = 4000, so 4000 ÷ 1000 × 100 = 400
		ii	475	1	A	Increase = 5700, so 5700 ÷ 12 = 475
		iii	7 months	1	C	Lower value given in stem of question was 5000.
	c		To provide information about cell processes/cell growth/differentiation/gene regulation **OR** Used as model cells to study how diseases develop **OR** Used as model cells for drug testing	1	C	Answers involving research only here. Cannot include therapeutic uses such as repair or replacement of damaged or diseased organs and tissues.
4	a		96	1	C	Note that the scale is 2 for each division along the x-axis.
	b		At 0 minutes/start the rate of urine production was 2 cm³/minute Increased to 20 cm³/minute at 90 minutes Then decreased to 2 cm³/minute at 140 minutes Then remained constant (to 180 minutes) **(All 4 = 2, 3 correct = 1, units required only once)** **(All figures correct but no units = 1)**	2	AA	Take care with these double y-axis graphs. Select the correct line and read the values from the correct y-axis. When answering this type of question, it is essential that you quote the values of the appropriate points and use the exact labels given on the axes in your answer. You must also use the correct units in your description.

Question		Expected answer	Mark	Demand	Commentary with hints and tips
	c	2 cm³ per minute	1	A	The maximum salt concentration is 0.6 g/100 cm³ which occurs at either 0 cm³ or 180 cm³. Reading down to the other graph at these points shows that the rate of urine production is 2 cm³ per minute.
	d	13.5	1	A	This requires you to use a ruler and carefully read the values from each graph, remembering to use the correct scale, then subtract one from the other.
	e	Analysis of an individual's genome/knowledge of the genetic component of risk of disease	1	A	Personalised medical treatment depends on knowing the individual's genome.
5	A	**1** There is variation of the phenotype **2** Stabilising selection favours the average phenotype **3** Stabilising selection selects against the extremes of the variation **4** The frequency of the average phenotype increases **(Labelled diagram acceptable)** **(All 4)**	4	CCCA	If you had a problem with this question, it would be useful to try writing the answers a few times to help you memorise the key points – the vocabulary is crucial.
	B	**1** There is variation of the phenotype **2** Directional selection favours one extreme of the variation **3** Directional selection selects against the other extreme/average phenotypes **4** The frequency of the extreme phenotype increases **(Labelled diagram acceptable)** **(All 4)**	4	CCCA	
6	a	**1** Sequence data = **1** **2** Fossil evidence = **1**	2	CC	**1** Data includes sequences of nucleotides/bases and amino acid sequences in specific proteins. **2** Fossils help to confirm accurate dating.
	b	680	1	C	980 − 300 = 680
	c	360	1	C	Remember that common ancestors are found at the T junctions of the phylogenetic tree.
	d	4	1	C	Remember, it's the number of animals which **share** the ancestor with the zebrafish so don't count the zebrafish itself.

	Question	Expected answer	Mark	Demand	Commentary with hints and tips
7	a	NAD **OR** FAD	1	C	These are the two coenzymes that carry the hydrogen ions and electrons to the electron transport chain.
	b	Phospholipid	1	C	The membrane has a phospholipid bilayer and patchy arrangement of proteins.
	c	Provide the energy for/are used to pump hydrogen ions across the membrane	1	A	The hydrogen ions are being pumped against their concentration gradient – this requires energy.
	d	ATP synthase	1	C	Whether the question asks you to name the 'protein' or the 'enzyme' in the membrane responsible for ATP production, the answer is ATP synthase.
	e	Final acceptor of hydrogen ions = **1** and electrons = **1**	2	CA	Remember to learn this role of oxygen.
8	a	Volume of extract/solution/ sample pH of solutions Time left in colorimeter/out of bath **(Any 2, 1 mark each)**	2	CA	Read the information and the experimental set up and note any conditions or quantities already mentioned. Choose the key variables that might affect the experiment that have not already been mentioned.
	b	Some reaction occurred/ enzyme started working/ browning occurred/pigment produced **AND** Immediately before lead ethanoate added/before reading taken/as soon as cut/ while being cut up **OR** Banana tissue was already damaged	1	A	This question requires you to realise that the damage to the tissue and the enzyme action will have started before the lead ethanoate is even added.
	c	Have tubes set up in exactly the same way, but containing sample of extract + water instead of lead ethanoate	1	A	The control should be identical to the original experiment apart from the one factor being investigated. If you are asked to describe a suitable control, make sure you describe it in full. A control experiment allows a comparison to be made and allows you to relate the dependent variable to the independent one.

Question			Expected answer	Mark	Demand	Commentary with hints and tips
	d	i	Appropriate enclosed scales with zeros and labels exactly from table **AND** Shared zero exactly at origin	1	C	The graph labels should be identical to the table headings, including units. Copy them exactly, leaving nothing out. Choose scales that use at least half of the graph grid provided, otherwise a mark will be deducted. The values of the divisions on the scales you choose should allow you to plot all points accurately. Make sure your scales include zero if appropriate and extend beyond the highest data points. Label the x-axis time (minutes) and y-axis colorimeter reading (units).
		ii	Points plotted accurately and joined with straight lines **AND** Line labelled from table	1	A	The points in a line graph should be joined with straight lines. Note in this graph you must label the new line you plot.
	e		As concentration/it increases, the activity (of the enzyme) decreases/effect of enzyme decreases/the enzyme is inhibited more **OR** As concentration/it increases, the rate of browning decreases **OR** Converse	1	C	This requires you to make a comparison between the different concentrations and the effect that they had on the browning/enzyme action.
	f		Repeat the experiment (many times) with both concentrations of lead ethanoate	1	C	Reliability is improved by repeating to boost confidence in the results.
9	a		Blood vessel – constricts to reduce flow of blood to skin surface so less heat is lost by radiation **OR** Sweat gland – reduced sweat production so less heat is lost by evaporation **OR** Hair erector muscle – contracts, raising hair to trap a layer of insulating air	1	A	Candidates often lose marks in these questions by failing to give full explanations. Make sure you can give these responses.
	b		Optimal temperature for enzyme-controlled reaction rates = **1** **AND** Optimal diffusion rates = **1**	2	CA	Most questions involving 'temperature' relate to enzymes.

Question			Expected answer	Mark	Demand	Commentary with hints and tips
	c		Conformers	1	C	Conformers do not use physiological mechanisms that require energy to alter their metabolic rate and so have a low metabolic energy cost. They have narrow ecological niches.
10	a	i	M and K	1	A	First, look at the chart and work out the scale and the positions of the fragments. The values are 54 and 46. This means that the enzymes must have cut the plasmid at positions M and K.
		ii	100 kb	1	A	One cut would give a fragment made up of the entire plasmid which is 100 kb long.
	b		(DNA) Ligase	1	C	Joins the complementary 'sticky ends' of the required gene with the 'sticky ends' of the plasmid. The recombinant plasmid acts as a genetic vector, carrying genetic material from one cell to another.
	c		Polypeptides can be folded incorrectly/may not be functional **OR** The bacteria cannot carry out folding of the protein	1	C	This problem can be overcome by using yeast instead of bacteria. Yeast is a special eukaryotic organism and so has the genes necessary to carry out the folding needed to make the protein functional. This is often poorly understood or explained by candidates. Take time to learn this.
	d		Mutagenesis **AND** Exposing the bacteria to UV light/other forms of radiation/X-rays/mutagenic chemicals	1	A	Mutagenesis is the process of inducing mutations, some of which might produce an improved strain with desirable qualities.
11	a	i	To eliminate any competition from contaminating microorganisms	1	C	The presence of other microorganisms could provide competition for the available resources.
		ii	Condition: Temperature/ oxygen/pH = **1** Description (must match): Temperature – cooling water/ jacket/thermostat Oxygen – aerator/sparger/ pump air in pH – add acid/alkali/buffer = **1**	2	CA	Remember **stop** = **s**terile conditions, **t**emperature, **o**xygen and **p**H – these are the controlled conditions for most bioreactors.

	Question		Expected answer	Mark	Demand	Commentary with hints and tips
		iii	0.5	1	C	These 'average increase' questions are often poorly done by candidates. First calculate the increase over the time period. Then divide it by the number of hours. $30 \div 60 = 0.5$
		iv	Will start to reduce **AND** this is because nutrients will start to run out **OR** toxins will build up	1	C	The fermenter is sealed so it is difficult to add more nutrient or to remove any toxic waste products the culture produces.
	b		Antibiotics = **1** Inhibit the growth of other species of bacteria and so reduce competition for available resources = **1**	2	CA	Remember that secondary metabolites are produced during the **s**tationary phase. **SS** – **s**tationary and **s**econdary.
12	a	i	The later the species departs, the slower the migration rate	1	C	Look for the trend on the graph – it's fairly straightforward.
		ii	Those species wintering in tropical areas depart earlier **AND** have higher rates of migration	1	C	Again, quite straightforward.
		iii	They have a longer journey to make	1	A	Need to use information from the stem which confirms that the tropical migrants travel further.
	b	i	Ringing/marking **OR** Satellite (GPS) tagging	1	C	These are the methods mentioned in the course specification – there are other marking methods.
		ii	Hibernation **OR** Aestivation **OR** Daily torpor	1	C	Toleration means being adapted to survive within the region of the adverse conditions.
13	a		May be free of predators/ parasites/pathogens/ competitors/May be able to outcompete native species	1	C	It is useful to know some of the reasons why introduced species are sometimes able to outcompete native species.
	b	i	Insecticide only – the population of fire ants decreased by 90%/fell to 10% one year after treatment then started to increase again = **1** Integrated approach – the population of fire ants decreased by 95%/fell to 5% one year after treatment and then remained constant = **1**	2	CA	You must provide a full description and include values and units if appropriate in these questions.

Question			Expected answer	Mark	Demand	Commentary with hints and tips
		ii	Some fire ants/those with resistance to the insecticide survive = **1** They reproduce and pass on their resistance/favourable/ beneficial gene to their offspring = **1**	2	CA	Be able to give a good description of the process of natural selection using the correct key terms and apply it to any new example given.
		iii	Introduced predator may become invasive/affect other native organisms/disrupt food web/reduce biodiversity	1	C	Typical KU question that you could convert into a question and answer card.
		iv	The scuttle fly larvae/parasite benefits in terms of energy/ nutrients whereas the host/fire ant is harmed	1	A	This is a +/− relationship. The parasite gains in terms of nutrients/ energy and the host loses in terms of loss of nutrients/energy.
14	a		To eliminate genetic differences which might have an effect on the results **OR** So that temperature was the only factor affecting results	1	A	Designed to make you consider the variables that need to be considered and controlled to allow valid conclusions to be made.
	b	i	As the temperature increased, the mean growth rate increased from 540 g/day at 0 °C to 850 g/day at 20 °C = **1** As temperature increased above 20 °C the mean growth rate decreased to 310 g/day at 45 °C = **1**	2	CC	Give full descriptions and include values and units if possible.
		ii	5:6	1	C	Quite a tricky ratio question. 450:540 can be simplified by dividing by 100 first. This gives 45:54. Then dividing by 9 gives 5:6.
		iii	45	1	C	Need to look at the data in the table and find the mid-point between the 10 °C and 20 °C values.
	c		Energy lost at each trophic level = **1** Eating crops removes a trophic level for the human food chain compared to eating animals = **1**	2	CA	This is knowledge straight from the learning outcomes in the course specification.
	d		Stereotypy/misdirected behaviour/failure in sexual behaviour/failure in parental behaviour/altered levels of activity/apathy/hysteria	1	C	These are the examples of poor livestock welfare that you are expected to remember.

	Question		Expected answer	Mark	Demand	Commentary with hints and tips
15	a		To ensure that they are homozygous/all show desired characteristics	1	C	Inbreeding gives a pure bred breed, homozygosity and uniform phenotype.
	b		To combine desired characteristic of both parent breeds **OR** To produce increased desirability/heterosis	1	C	The offspring combine the desired parental phenotypes and may also have other improvements such as higher yield or disease resistance.
	c		F_1 hybrids would produce offspring which were too varied **OR** Not guaranteed to show the desired characteristics	1	A	Because the hybrids are heterozygous, their offspring would be varied and unreliable for commercial production.
	d		To ensure that the offspring will show these characteristics	1	C	Homozygous dominant individuals always pass their phenotypic characteristics to their offspring.
16	A	i	**1** Social hierarchy is a rank order/pecking order in a group of animals **OR** Dominant/alpha **AND** subordinates/lower rank **2** Aggression/fighting/ conflict/violence reduced **3** Ritualistic display/ appeasement/threat/ submissive behaviour **4** Alliances formed to increase social status **OR** Descriptive examples **5** Ensures best/successful genes/favourable alleles/ characteristics are passed on **6** Guarantees experienced leadership **(Any 4)**	4	CCCA	Once again, a large number of key terms and definitions are required. Flash cards produced and memorised for each key area are an effective method of achieving high scores in many extended response questions.

Question			Expected answer	Mark	Demand	Commentary with hints and tips
		ii	**1** Co-operative hunting is where animals hunt together to obtain food	5	CCCCA	
			2 Increases hunting success			
			OR			
			Allows larger prey to be brought down/hunted and killed			
			OR			
			More successful than hunting individually			
			3 Subordinate animals all get more food/energy than hunting alone			
			OR			
			Less energy used/lost per individual			
			4 Social defence strategies increase the chances of survival as some individuals can watch for predators while others can forage for food			
			5 In social defence, groups may adopt specialised formations when under attack, protecting their young			
			OR			
			Example such as wild oxen form ring with horned males on outside			
			6 Early warning of predators can be given			
			OR			
			Example such as bird alarm calls/meerkat lookouts			
			7 Predators intimidated/confused			
			OR			
			Example such as stampeding zebra can confuse predators/several crows mobbing a single buzzard			
			(Any 5)			

Question			Expected answer	Mark	Demand	Commentary with hints and tips
	B	i	**1** Light energy excites electrons in pigment molecules **2** High energy electrons pass through an electron transport chain **3** Releasing energy to generate ATP **4** By ATP synthase **5** Energy also used to split water into oxygen which is released **6** And hydrogen which is transferred to the coenzyme NADP **(Any 5)**	5	CCCCA	An in-depth understanding of the stages in photosynthesis is advisable for both short-answer structured questions and extended response questions such as this. A good, well-labelled diagram is an excellent way to present this information and will also help you to structure your written response. Practise drawing the light dependent stage and the Calvin cycle.
		ii	**1** Carbon dioxide joined to RuBP by RuBisCo **2** ATP used to phosphorylate an intermediate to form G3P **3** Hydrogen from NADPH used to form G3P **4** G3P forms sugars **5** G3P regenerates RuBP **(Any 4)**	4	CCCA	